A Colour Atlas of
THE NAIL
in Clinical Diagnosis

D. W. Beaven
MB CHB(NZ), FRACP, FRCP(Lond), FRCP(Ed), FACP(Hon.)

S. E. Brooks
(FNZIMBI)

Wolfe Medical Publications Limited

To Gwenda and Terry, who made the compilation of these illustrations possible, and to Dorothy, without whose valuable and dedicated assistance, this book would not have been possible.

Copyright © D. W. Beaven, S. E. Brooks 1984
Published by Wolfe Medical Publications Ltd 1984
Printed by Royal Smeets Offset b.v.,
Weert, Netherlands
ISBN 0 7234 0826 2

General Editor, Wolfe Medical Books
G. Barry Carruthers MD(Lond)

This book is one of the titles in the series of Wolfe Medical Atlases, a series which brings together probably the world's largest systematic published collection of diagnostic colour photographs.

For a full list of Atlases in the series, plus forthcoming titles and details of our surgical, dental and veterinary Atlases, please write to Wolfe Medical Publications Ltd, Wolfe House, 3 Conway Street, London W1P 6HE.

Contents

Acknowledgements

We have been much indebted to many friends and helpers over 20 years.

Sally Collins, head nurse of the Professorial Medical Unit at the Princess Margaret Hospital, whose joint interest 'collected' many of these patients from the general medical and endocrine services.

Our dermatological colleagues in Christchurch, Allan Muir and Derek Larnder, have most generously lent a wide range of their valued slides from teaching collections.

We value the co-operation of Mr Richard Winder, principal tutor of the New Zealand National Podiatry School who generously made available to us many of the original slides in the section on toenails.

Dr James Marshall of Templeton Hospital rendered invaluable help in collecting examples from his large group of intellectually handicapped patients.

Above all, we have been grateful to a series of professional colleagues in the Audio-Visual and Photographic Departments at Christchurch and the Princess Margaret Hospital, who assisted in so many ways; in particular, Mrs Fiona Van Oyen, medical illustrator at the Princess Margaret Hospital for her skill and help in completing the drawings.

Individual clinicians have lent slides from their undergraduate teaching collections: Drs Thornley, McLeod, Bailey, Swainson, Janus, Moller, Brownlie, Ikram and Mr Greg Coyle filled in gaps in various sections.

To those many helpful people, departmental and other hospital staff members and colleagues, and others too numerous to mention by name, we remain greatly in their debt.

Above all, without the willing help of all those whose nails appear herein, this book would not have been possible.

Preface

Looking carefully at the nails should be an essential prerequisite in the examination of a patient. This colour atlas we hope will illustrate the many and varied changes which can occur.

It is the belief of the authors and publishers that such a valuable site for observation of health, personality and general medical disorders as the nails has been much neglected. Other nail books have generally been written by dermatologists or hand specialists. There is a current world-wide swing towards making more use of physical examination when constructing problem lists about patients. We hope that this volume of *mainly common* nail changes will therefore be useful to students, general practitioners, district nurses and podiatrists.

Conditions in the nails frequently seen when first introducing oneself to the patient may lead to subsequent examination or follow-up. The changes seen in these illustrations represent 99 per cent of abnormal findings seen by family physicians, specialists and resident hospital staff.

Hitherto, changes in the nails have often been regarded as due to age. Normal older people have normal nails; we hope these photographs will help identify those who do not!

Chapter 1

The Normal Nail

The fingernails and toenails have remained through a long evolutionary period. They protect the sensitivity of the tips of the fingers and toes and are used for offensive and defensive activities. In the hands, the nails increase the fine function of the fingertips.

The nail plate is made up of specialised keratin made in the cells of the nail bed. An invagination of the epidermis forms these nail-producing organs which are fully formed halfway through embryo life. The nail is made up of the nail plate, the nail bed, with the lunula and the nail folds.

The nail plate is the largest sheet of keratin in the body and is of crucial importance in clinical medicine because it reflects the health events of the previous months. The nail plate is formed from matrix cells of the nail bed. Some contribution takes place to the forward growth from the cells under the proximal nail fold. A very detailed description can be found in Lewin (1965) of the three embryological elements – dorsal, intermediate and ventral – which go to make the whole.

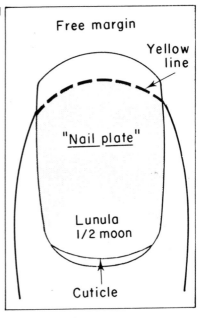

1

The nail plate grows slowly forward until it breaks free of the nail bed at the yellow line. The nail plate beyond the yellow line leads to the free edge which is of variable length and thickness, depending upon frequency of cutting and occupation. The area under the free edge is called the hyponychium.

The nail bed consists of a deeper portion or dermis and the outer section or epidermis. The section of the nail bed under the proximal nail fold, the cuticle and the lunula, is also often referred to as the matrix. The dermis of the nail bed is arranged in longitudinal grooves and longitudinal ridges. The epidermis is thickened and gradually passes into the nail with swelling of matrix cells, nucleolysis

Free margin

Yellow line

"Nail plate"

Lunula 1/2 moon

Cuticle

1 Normal nail.

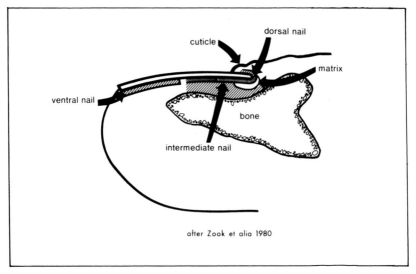

cuticle | dorsal nail

matrix

ventral nail

bone

intermediate nail

after Zook et alia 1980

2

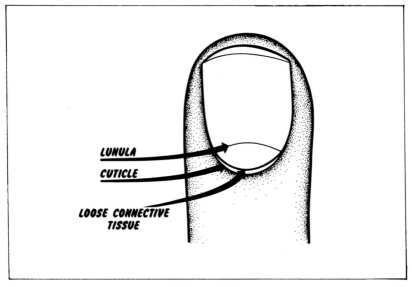

LUNULA

CUTICLE

LOOSE CONNECTIVE
TISSUE

3

and subsequent shrinkage. The lunula or white 'half moon' at the base or proximal end of the nail is particularly smooth, flat and shiny. The whiteness of the lunula is still a matter of controversy but its absence is notable and important. In certain chromosome abnormalities the lunula is absent, i.e., monosomia 4 and it may be diminished in trisomy 21.

At the edges of the nail the skin turns into the nail folds and this junctional area between skin and nail is known as the nail groove. The outer end of the nail groove is important in the great toe because of the frequency of 'in-growing'.

At the proximal end of the nail is the eponychium with its loose free border or cuticle.

Studies on the composition of the nails show moderately high concentrations of sulphur, selenium, calcium and potassium, but trace element losses from nails are small.

References

1. Forslind, B. On the structure of the normal nail. A scanning electron microscope study. *Archives fur Dermatologishe Forschung*, 1975, **251**, 199.
2. Harrison, W. W. & Clemena, G. G. Survey analysis of trace elements in human finger nails by spark source mass spectrometry. *Clinica Chimica Acta*, 1972, **36**, 485.
3. Lewin, K. The normal finger nail. *British Journal of Dermotology*, 1961, **77**, 421.
4. Vellar, O. D. Composition of human nail substance. *American Journal of Clinical Nutrition*, 1970, **23**, 1272.
5. Runne, U. The human nail: structure, growth and pathological changes. *Current Problems in Dermatology*, 1981, **9**, 102.
6. Zook, E. G., Van Beek, A. L., Russell, R. C. & Beatty, M. E. Anatomy and physiology of the perionychium: a review of the literature and anatomic study. *Journal of Hand Surgery*, 1980, **5**(6), 528.

Blood supply

The nail bed is well supplied by small branches from dorsal and ventral anastomoses. The two digital arteries run on the palmar aspect or ventral side of the finger. They give off small branches just prior to and after the terminal digital joint and then form a rich cruciate anastomosis on the palmar aspect of the terminal pulp.

Here are the very important arterio-venous capillary shunts partially under nervous control and partially controlled by circulating kinins. The physiology of these digital capillary shunts allows blood to proceed through the fingertip pulp and return to the nail matrix area by two dorsal arteries; this has been well demonstrated by Flint (1955). These capillary shunts terminate in a further dorsal anastomosis which also receives blood from the branches of the digital arteries mentioned before.

Thus the base of the nail bed and matrix cells can, when the terminal capillary pulps are open, receive a double blood supply with relative increased activity of the cells. This accounts for the sponginess seen in clubbing of the fingernails. Because of these collateral vessels near the fingertip, the finger pulp and the nail beds seldom show much ischaemia.

However, the two digital arteries run on the sides of the fingers as do the returning dorsal arteries from the finger pulp. This means that earliest changes in the blood supply or nutrients to the nails will be seen first in the centre of the nail at the base, or proximal part of the nail.

The central portion of the nail bed may also be involved and as nail thickness is added from the ventral surface well down the nail, such central lesions from blood supply or nutrient deficiency will become exaggerated towards the free edge of the nail.

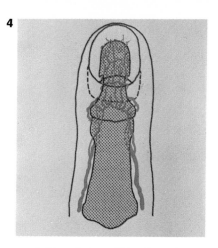

4 Dorsal view of blood supply.

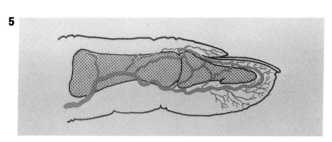

5 Lateral view of blood supply. Adapted from Flint. Note supply of matrix via capillary fingertip shunt.

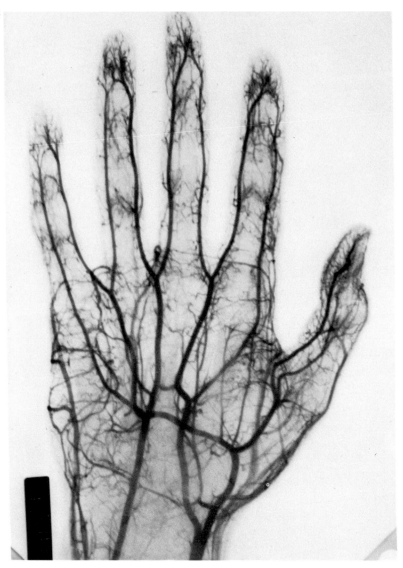

6 Angiogram of normal hands showing arterial plexus at fingertips.

7

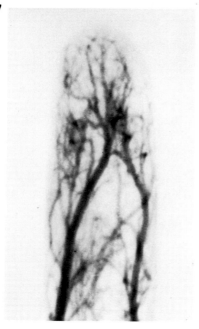

7 Close-up fingertip and nail bed plexus normal. These come from a normal hand and normal angiogram.

References

1. Flint, M. H. Some observations on the vascular supply of the nail bed and terminal segments of the finger. *British Journal of Plastic Surgery*, 1955, 8, 186.
2. Norton, L. A. Incorporation of the thymidine-methyl H^3 and glycine 2 H^3 in the nail matrix and bed of humans. *Journal of Investigative Dermatology*, 1971, 56, 61.
3. Samman, P. D. Abnormalities of the finger nails associated with impaired peripheral blood supply. *British Journal of Dermatology*, 1962, 74, 165.
4. Walters, K. A. Physiochemical characterization of the human nail: I. pressure sealed apparatus for measuring nail plate permeabilities. *Journal of Investigative Dermatology*, 1981, 7(2), 76.
5. Zook, E. G. Anatomy and physiology of the perionychium: a review of the literature and anatomic study. *Journal of Hand Surgery*, 1980, 5(6), 528.

Method of examination

Urbanisation reduces the powers of natural observation. Once, man survived because of the deductions the tribal leaders derived from their observations.

All human beings carry the record of their personality and recent past health in a flat sheet of keratin of the nail plate. Careful observation of the nail plate and the infinite variations in colour, shape and care, may add greatly to the clinical or personal examination. Not only are there nearly always signs to observe in even healthy young people (such as habits, mannerisms and manicure) but increasingly minor abnormalities in the nutrition or health may be accumulated in the nail plate over recent months. As the thumbnail contains the largest area of nail, it may demonstrate most markedly any lesser changes. In examination of the nails the following should be noted.

Have a good light source, preferably sunlight. Light from an angle will reflect off the normally shiny surface of the nail plate. To observe this so-called 'shine' on the normal healthy nail is an important part of the proper examination of the fingernails. The fingers and nails may need to be moved up and down until the angle of reflected light is correct.

In order to be indicative of general disease or disorder and not merely due to local trauma, changes should be seen in all fingernails of the same hand.

Magnification may be needed, usually four or eight times, especially where the nails are small.

Biopsy of the nail folds is seldom useful but scrapings are valuable when infections are suspected. We have seen ill-judged biopsy of the nail bed lead to long-term nail disfiguration.

We advocate a fast but careful examination in every person presenting with symptoms of illness. To do this we suggest 'building-in' to the normal greeting and pulse-taking, the examination of the nails. In the sequence which follows a natural two-handed greeting leading to the left hand seeking the radial pulse at the wrist, the examiner's right hand may lightly hold the fingers. After observations for warmth, colour and shape, the hand is turned over and the fingers flexed upwards or downwards until the light source reflects on the shiny nail plate.

It is only by such a standardised but simple procedure that early and minor changes may be detected. Examination of the nails should be part of the full clinical examination of every patient and in poorly lighted areas a lighted magnifying lens may be used.

A standard technique as set out here allows the doctor, nurse or other health worker, quickly to determine the limits of colour or normality for any ethnic group. This much neglected and useful clinical aid is thus easily transportable to all countries without technical, cultural or language barriers.

Approach and examination of the nails
This next series of figures shows the normal approach to an ambulant patient.

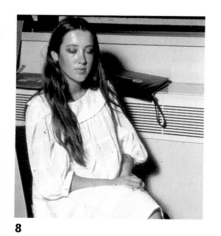

8

8 Greeting from door.

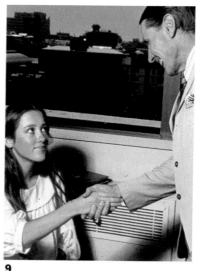

9

9 Greeting with handshake.

10

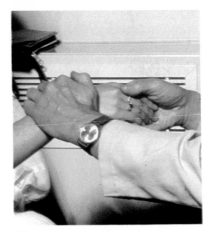

11

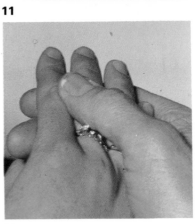

12

13

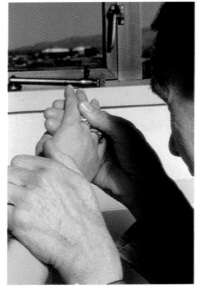

14

15

10 Normal approach to sit down and shake hands whilst having a social chat.

11 The normal two-handed greeting with the examiner's left hand palpating the slightly flexed wrist. The right hand initially holds the fingertips to determine warmth, colour, circulation, etc.

12 The hand is next swung up so the nails come into view whilst the pulse continues to be felt.

13 The observer now lines up his eye along the line of the nail and the daylight from the window. The normal nail should show a characteristic 'shine' from reflected light if held at the proper angle.

14 This angle should be adjusted until a bright light is reflected off the full nail length, thus highlighting abnormalities.

15 Several nails should always be studied although the thumbnail or great toenail, being the largest sheet of nail, usually shows the most marked changes. To be significant any changes should be shown in *all* nails on that hand.

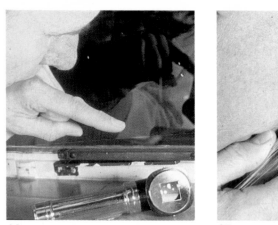
16

17

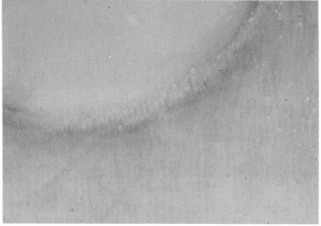
18

16 Method of examination with good sunlight shining down the nail and observer behind and above, note also lighted hand lens (lying on table) for closer inspection.

17 Examination of the nail using a lighted magnifying speculum when in poor light.

18 Edge of normal eponychium or proximal nail fold magnified.

References

Biopsy of the nail

1. Pierre, M. *The Nail*. Churchill-Livingstone, Edinburgh, 1981, p.3.
2. Scher, R. K. Punch biopsies of nails: a simple, valuable procedure. *Journal of Dermatological and Surgical Oncology*, 1978, **4**(7), 528.
3. Scher, R. K. & Ackerman, A. B. Subtle clues to diagnosis from biopsies of nails. The value of nail biopsy for demonstrating fungi not demonstrable by microbiologic techniques. *American Journal of Dermatopathology*, 1980, **2**(1), 55.
4. Stone, O, J., Barr, R. J. & Herten, R. J. Biopsy of the nail area. *Cutis*, 1978, **21**(2), 257.

Growth

The continuous growth of the nails throughout life proceeds at a slower pace in old age. The growth from the base of the nail to the free edge is rapid in small children (6 to 8 weeks) but in normal adults varies between 0.5 and 1.2 mm/week.

Because significant growth may be seen each day, a few days of serious illness gives rise to that most helpful of clinical signs in the nails – the growth arrest line, which is discussed under the heading 'Beau's Lines'. William Bean measured his own left thumbnail growth each day over 30 years and found this to be 0.123 mm/day at 32 years of age and 0.100 mm/day when aged 61.

Temperature probably has some effect as earlier polar observations showed slowed growth rates, but these are now the same as in temperate climates (Donovan).

As the nails grow faster on longer fingers, the fastest growth occurs in the middle finger, followed by the ring and index finger. Hamilton *et al.* (1955) in their classic study showed that growth rate was most significantly related to ageing. Whilst the nails grew more slowly in later decades, they were thicker. Thus, in old people with normal digital blood supply, the same unit mass of nail is probably formed each day.

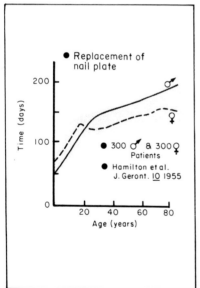

More recently, studies by Griffiths & Reshad (1983) have confirmed the previous observations that trauma stimulates growth. The nails of the right hand grow faster in right-handed people, the dominant hand is used more often, thus being subject to more minor trauma.

Dawber (1981) also confirmed that immobility can slow up the rate of nail growth. Some rare and unusual conditions such as the yellow nail syndrome described by Samman also lead to very slow growth. The reason that slowed growth occurs in paralysed fingers has not been explained, but may be due to lack of stimulus and minor trauma.

19

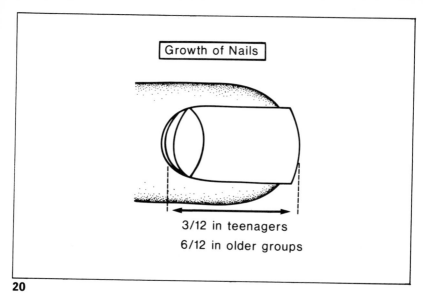

20 Approximate guide to growth rates. 6 to 8 months for full re-growth would be seen in 70-year-olds.

SLOW NAIL GROWTH

- Age
- Genetic
- ♀s

not: hair
    baldness
    fat
    height & weight

21

FAST NAIL GROWTH

- Nail biting
- Damage nail tip
- 1st finger

22

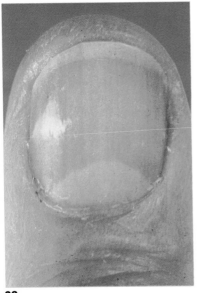

23

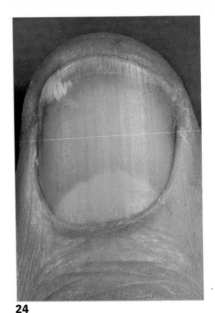

24

23 1st photograph of thumb.

24 2nd photograph 6 weeks later.

Nail growth in 58-year-old physician over 6 weeks as measured by movement of 'white spot' resulting from previous injury.

By far the most important factor about growth is the known 3 to 4 months time for an arrest line to travel from the base to the free edge. Thus, knowing the individual's age, one can calculate the likely date of a previous serious illness.

References

1. Baran, R. Nail growth direction revisited. Why do nails grow out instead of up? *Journal of the American Academy of Dermatology*, 1981, **4**(1), 78.
2. Bean, W. B. Nail growth 30 years of observation. *Archives of Internal Medicine*, 1974, **134**, 479.
3. Dawber, R. The effect of immobilization on fingernail growth. *Clinical and Experimental Dermatology*, 1981, **6**(5), 533.
4. Donovan, K. M. Antarctic environment and nail growth. *British Journal of Dermatology*, 1977, **96**(5), 507.
5. Griffiths, W. A. & Reshad, H. Hair and nail growth: an investigation of the role of left- and right-handedness. *Clinical and Experimental Dermatology*, 1983, **8**(2), 129.
6. Hamilton, J. B., Terada, H. & Mestler, G. E. Studies of growth throughout the life span in Japanese; growth and size of nails and their relationship to age, sex, heredity and other factors. *Journal of Gerontology*, 1955, **10**, 401.
7. Orentreich, N., Markofsky, J. & Vogelman, J. H. The effect of ageing on the rate of linear nail growth. *Journal of Investigative Dermatology*, 1979, **73**(1), 126,

Normals

There are some variations seen in the nail appearances which relate principally to the habits and occupation of the person being examined.

The nail can react in only a few ways if the nail-forming cells or their blood supply or nutrition are impaired. Likewise, distal injuries almost always arise as a result of infections of a bacterial or fungal nature ascending from the free edge.

Insufficient scientific studies have been carried out on large enough groups of subjects to ascertain whether dietary factors may influence texture and appearance. Studies on flexibility and hardness of the nails (Ramrakhiani, 1978; Finlay *et al.*, 1980) suggest that these may influence occupational response. Likewise, Forslind *et al.* (1980) have also reported that detergents, organic chemicals, oils, etc. may be factors in the occupational environment which affect the response.

A group of people in various occupations are presented to describe the wide limits of normality found in persons who otherwise seem to be in good general health.

In the general examination of the patient we believe that signs of occupational activity may allow the examiner to establish a quicker rapport than would otherwise be the case.

25 Normal first or index fingernail in a young woman.

26 Normal nail of a young woman in her 20s.

27 Normal thumbnail of a 27-year-old female scientific officer. ? Hint of Beau's lines.

28 Normal thumbnail of a young female laboratory assistant.

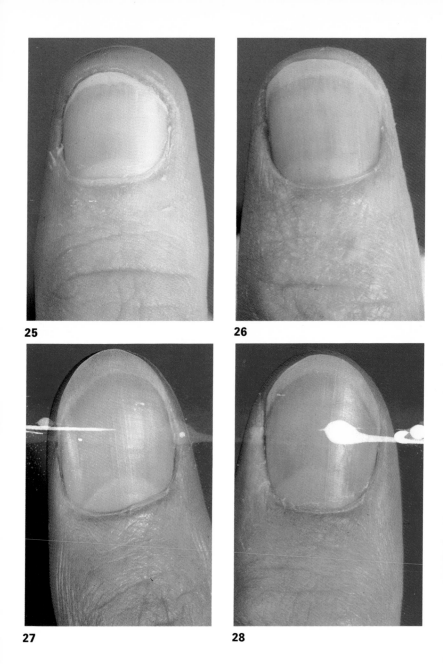

25

26

27

28

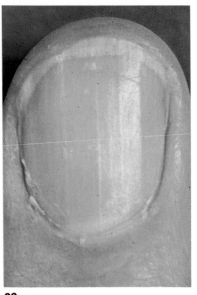

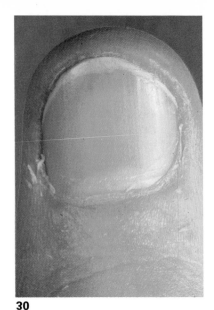

29 **30**

29 Normal thumbnail of a young female laboratory assistant.

30 Normal young female clerical assistant.

31 The normal thumbnail from a healthy female laboratory worker in her 20s.

32 Normal thumbnail of a young female laboratory assistant.

33 Normal thumbnail of a 22-year-old laboratory technician.

34 Thumbnail from normal healthy female laboratory worker in her 20s.

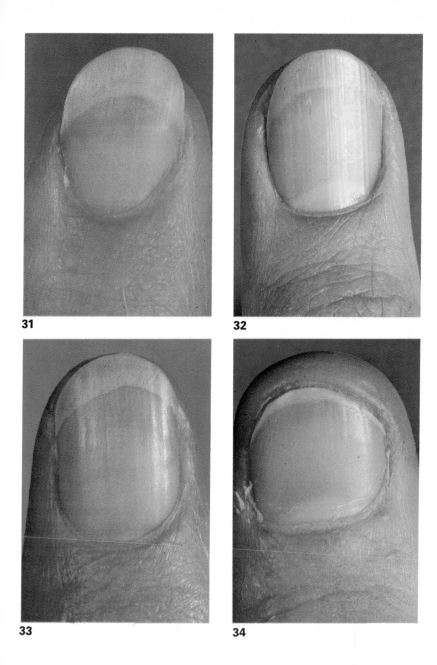

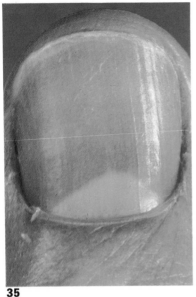
35

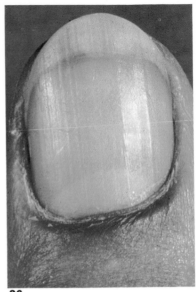
36

35 Normal young female receptionist.

36 Thumbnail from another young female laboratory worker.

37 Thumbnail from normal young laboratory technician.

38 Normal thumbnail of a 27-year-old male clerical worker.

39 Normal thumbnail in a 30-year-old male clerical worker.

40 Occupational injury to the edge of nail and also the skin in a 35-year-old electrician.

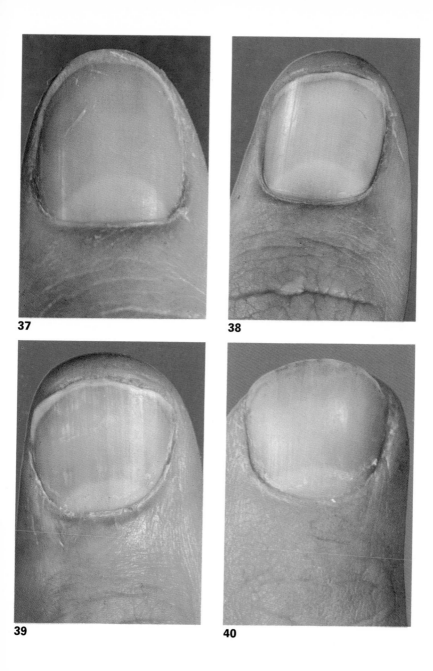

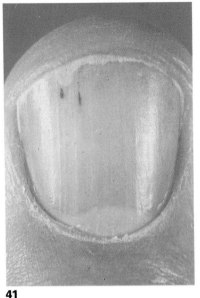 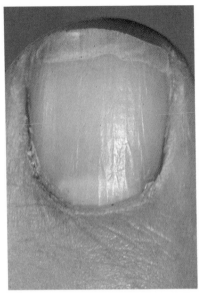

41   **42**

41 Normal thumbnail of young fitter and turner. The two 'splinter' haemorrhages and localised onycholysis are due to local trauma.

42 Thumbnail of a normal healthy young motor mechanic with no disability.

43 Motor mechanic. Note ingrained oil and occupational scratches on nail plate surface.

44 A middle-aged painter in normal health. Some onycholysis present together with occupational thickening of leading edge. Longitudinal ridges are a normal variation in this 55-year-old male.

45 Thumbnail from middle-aged house painter. Note signs of occupation and leading edge of nail used for 'flicking' open tins!

46 Thumbnail of more junior gardener. On the day of volunteering for photography the dirt-stained nail is trimmed right down. Note skin changes around the nail edge.

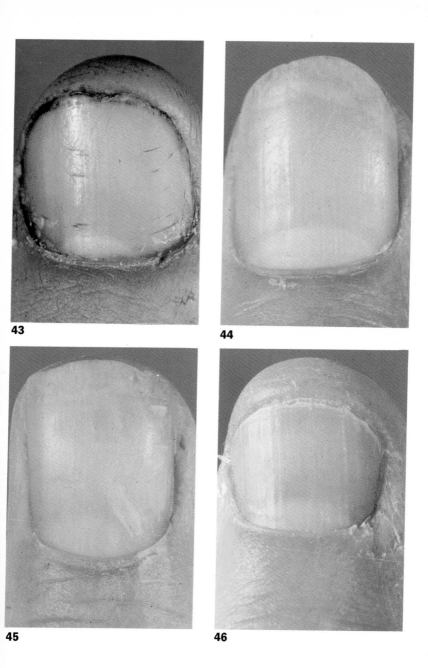

43

44

45

46

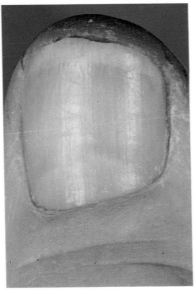 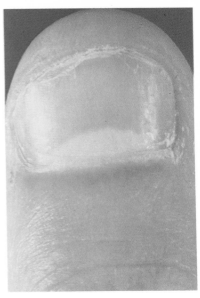

47 **48**

47 Normal thumbnail from healthy 40-year-old electrician. Minor occupational changes in nail.

48 Thumbnail from 26-year-old male orderly who is responsible for collecting and distributing crates of milk and oxygen cylinders. Note minor injury to nail from occupation.

49 Another normal thumbnail from a male hospital orderly in good health.

50 Thumbnail of hospital house manager, note small white spot of minor trauma 3 months ago on the free edge. Longitudinal ridges are a normal variation in middle life.

51 Thumbnail of 50-year-old head gardener in normal health. Despite care and attention to nails his occupation produces some early onycholysis.

52 Nail from youngish male in normal health, working in engineering workshop. As a fitter and turner he damages the leading edge which is cut short. Excessive manicure is shown on the right side of the nail.

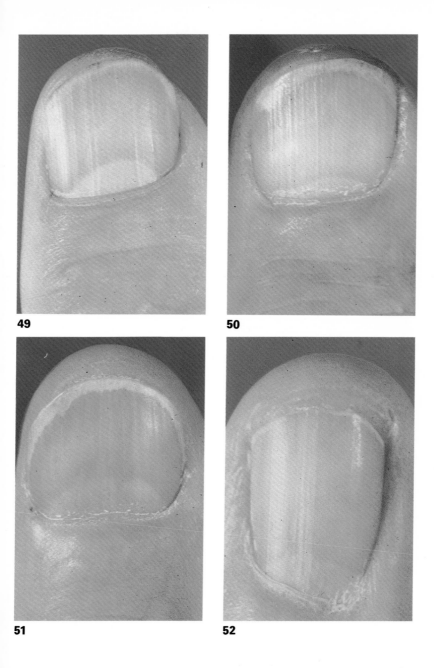

49

50

51

52

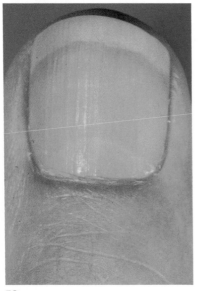 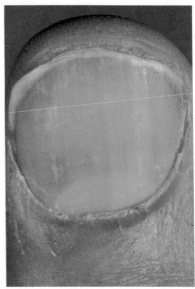

53　　　　　　　　　　　　　**54**

53 Normal thumbnail from a senior hospital orderly. Note the wide free edge.

54 Thumbnail from older female (in late 50s) who washes and packs glassware. Note white spots due to minor occupational injury and also longitudinal ridges.

55 Thumbnail from normal 62-year-old woman whose life-long hobby is working in her rock garden. Nails show occupational changes typical of active hard use.

56 Thumbnail from normal young female employed as washing-up assistant in laboratory.

57 Thumbnail from middle-aged female laboratory worker in good health.

58 Thumbnail from normal male laboratory worker. Note single longitudinal ridge.

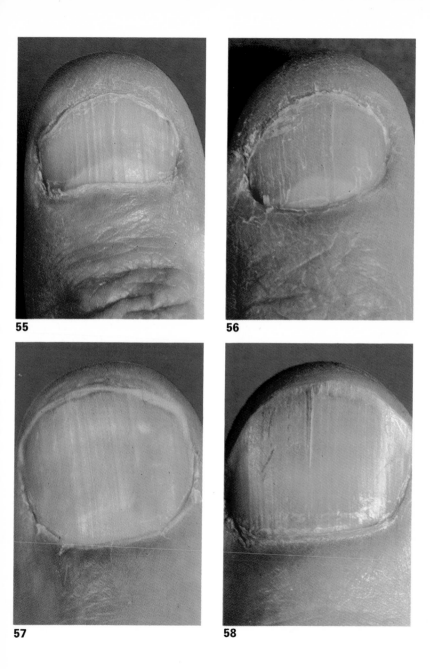

55

56

57

58

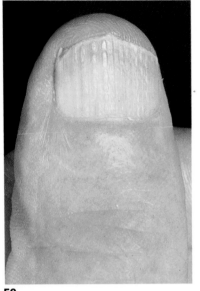 

59 **60**

59 Retired nail post office worker, 72 years of age, said to be in good health. Has however rather opaque nails and beading. ?Nutritional.

60 Fingernail of guitar-player. Note very short and typical thick pad of flesh over the tip of the finger.

References

1. Finlay, A. Y., Frost, P., Keith, A. D. & Snipes, W. An assessment of factors influencing flexibility of human fingernails. *British Journal of Dermatology*, 1980, **103**(4), 357.
2. Forslind, B., Nordstrom, G., Toijer, D. & Eriksson, K. The rigidity of human fingernails: a biophysical investigation on influencing physical parameters. *Acta Dermato-Venereologica (Stockh)*, 1980, **60**(3), 217.
3. Jarrett, A. & Spearman, R. I. C. Histochemistry of the human nail. *Archives of Dermatology*, 1966, **94**, 652.
4. Ramrakhiani, M. Indentation and hardness studies of human nails. *Indian Journal of Biochemistry and Biophysics*, 1978, **15**(4), 341.

Minor variations from normal

Thickening of the nail normally occurs with increasing age, so do longitudinal ridges on the surface. Such longitudinal ridges never normally occur before the age of 40 years and when present, suggest an associated chromosomal or congenital defect.

Increased curvature of the nail (onycholysis, koilonychia, etc.) is always associated with a more general systemic disorder. Many minor variations from normal are seen in association with other present or past skin diseases.

We have repeatedly noticed minor variations of the nails present in people with other general chronic medical illnesses. These have been included in the appropriate sections.

Marked longitudinal ridging, beading of the ridges and pitting of the nail plate do not occur in normal, young, healthy subjects. These appearances would suggest a more biochemical or medical search for otherwise unsuspected illness.

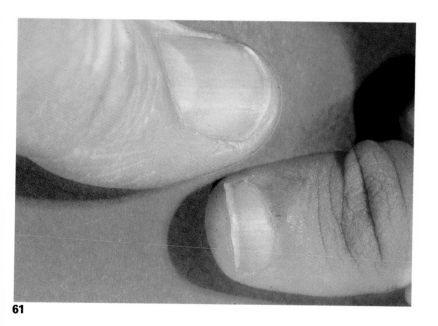

61

61 Small nail. Micronychia in intellectually impaired patient.

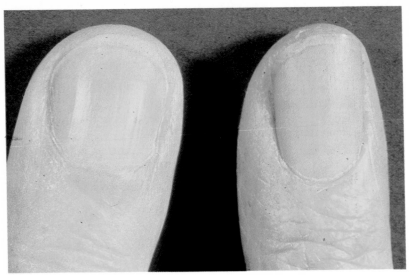

62

62 Variation in width. A normal variation in width and curvature seen in the left finger and not associated with any illness or known disease.

63 Triangular lunula. Pyramidal lunula in a trisomy 21 patient. Note also triangular cross section to the nail.

64 Triangular lunula. The triangular lunula in the thumb of obsessive female filing clerk, nail base, growth arrest lines and white streak suggest excessive manicure.

65 Small pits. These greenish coloured nails show bands of pigment but are shown here to illustrate pitting. Pitting of the nails is sometimes called Rosenau's Depressions. Psoriasis has been excluded. When widespread may produce so-called 'thimble-nail'.

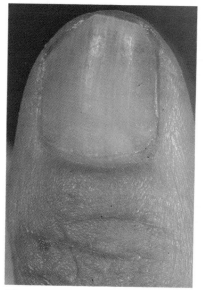

63

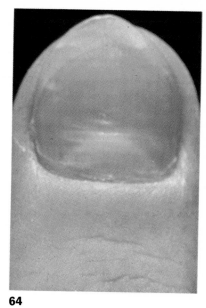

64

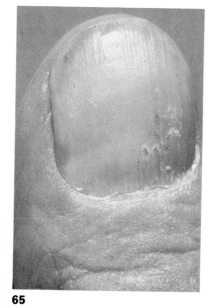

65

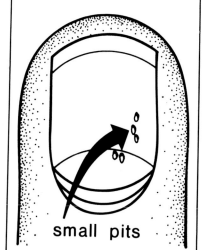

small pits

65a

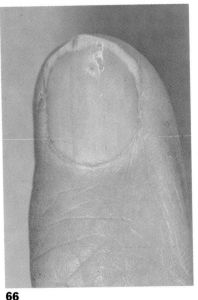
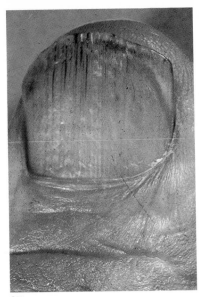

66   **67**

66 Minor pitting. Minor pitting in well-kept nails of a 29-year-old shop assistant, search well for 'hidden' psoriasis on elbows, knees or scalp! None found here.

67 Marked beading. Marked beading in an elderly and somewhat socially neglected pensioner with thyrotoxicosis.

68 Beading. Beading present in the longitudinal ridges of a middle-aged woman with poor manicure. Short nail may suggest some trisomy, and lack of lunula also sometimes congenital. Patient appeared of normal intelligence with no serious systemic disease.

69 Marked beading. 'Beading' on fingernail. Hypothyroid male on treatment with thyroxine for hypothyroidism. Note lack of beading near lunula and failure of care suggesting lapse in treatment.

70 Measurement of beading. 5th finger left hand on a small woman showing marked beading. Actual mm scale shows 'bead' to be 0.2–0.3 mm in length.

71 Ridges. Longitudinal ridges in a man with chronic obstructive airways disease and mild hypoxaemia ($pAO_2 = 70$).

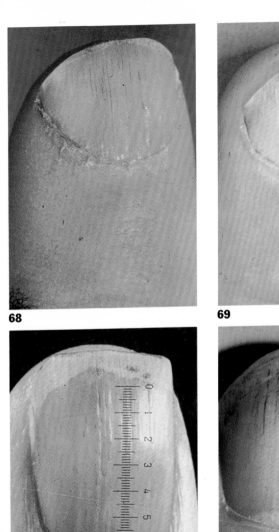

68

69

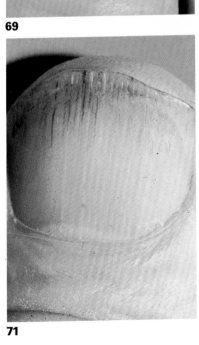

70

71

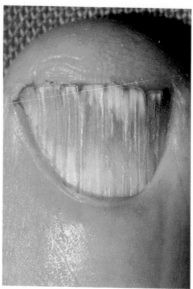

72 **73**

72 Very marked ridging. Longitudinal ridges in an elderly man with renal failure. Note 'pigment' in ridge is dirt from his hobby of rose-gardening.

73 Brachyonychia. Normal variation of short nail or brachyonychia. So-called 'racquet nail'. Sometimes seen with other congenital abnormalities. Look for toe changes, alterations in facial bones (Rubenstein & Taylor, 1963) and occasionally septal defects. Note also very marked and abnormal ridging and pigment band.

Care and personality

With repeated observations minor variations in the normal nail can reveal to the examining physician some features about the personality of the patient being examined. The care of the free edge, the state of manicure of the cuticle and the edges of the nail folds all reveal how well a person is in control of their immediate environment. Thus, from an early age, personality traits may result in habits such as nail- and cuticle-biting. Damage may also occur to the nail plate because of a habit tic. These habit tics may give rise to nail plate ridging and opacity, and the cuticle may pull away from the nail.

Occasionally, paronychia or fungal infections result from over-zealous manicure of the cuticle. This or nail-biting may result in shredding of the proximal nail fold. This is also called 'hang nail'. Unexplained transverse ridging found only on one nail may be caused by cuticle pressure from orange sticks or teeth.

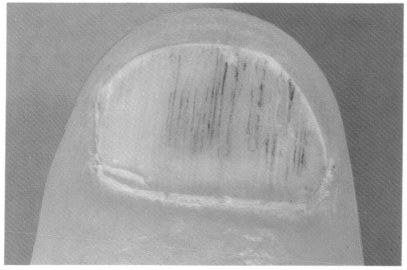

74

74 Many abnormalities. 64-year-old printing operator with chronic obstructive airways disease and auto-immune hypothyroidism. Has not worked for 6 weeks – note pigment (ink) line in nails. On treatment with thyroxine. Note also poor manicure.

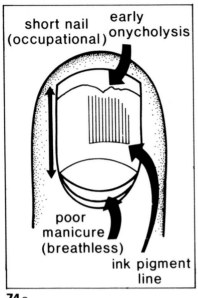

short nail (occupational)

early onycholysis

poor manicure (breathless)

ink pigment line

74 a

75

75 Poor manicure: why? Poor manicure as an indication of mental depression. The nail of a depressed widower in his late 60s. Note prominent longitudinal ridges.

76 Poor manicure. Poor trimming and cutting of nails with poor manicure in poorly educated motor car garage labourer.

77 Chemical damage. Dystrophy in a 36-year-old barman. This is most likely due to chemical damage from detergent powder used for washing glasses as small area of contact.

78 Leukonychia striata. Excessive manicure and pushing down of 'quick' – leads to white transverse injury lines – similar to leukonychia striata.

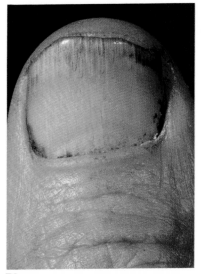

76

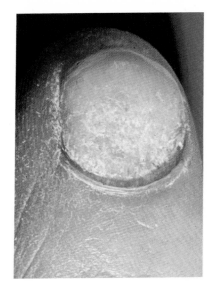

77

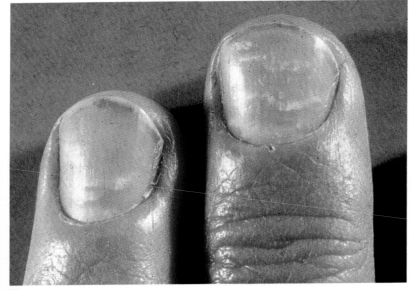

78

79

79 Camouflage. Normal nail, possibly slightly curved in a hospital receptionist. Well disguised! Danger of onycholysis.

80 Distressed housewife. Example of evidence of background and general medical conditions from various changes in the nails on one hand. Note filling in of nail fold due to mild bronchiectasis. Also paronychia of index fingernail suggesting immersion in water due to washing dishes or clothes.

81 Nail biting. Nail biting of all fingers. Brachyonychia or short nail has resulted in the middle finger.

References

1. Calnan, C. D. Reactions to artificial colouring materials. *Journal of the Society of Cosmetic Chemistry*, 1967, **18**, 215.
2. Samman, P. D. Onychia due to synthetic nail coverings. Experimental studies. *Transactions of the St John's Hospital, Dermatological Society London*, 1961, **46**, 68.
3. Samman, P. D. A traumatic nail dystrophy produced by a habit tic. *Archives of Dermatology*, 1963, **88**, 895.
4. Samman, P. D. Nail disorders caused by external influences. *Journal of the Society of Cosmetic Chemistry*, 1977, **28**, 351.

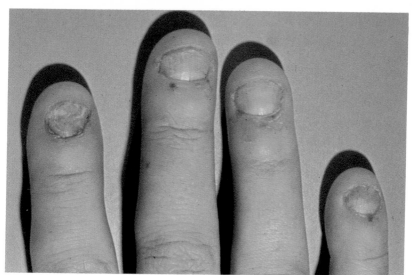

80

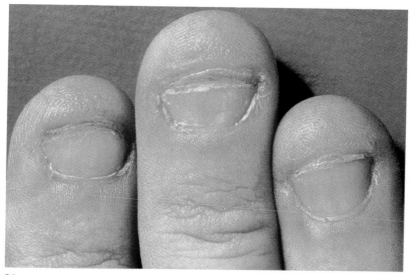

81

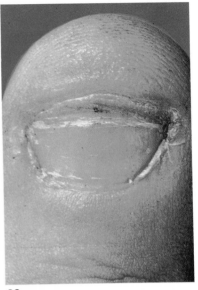

82

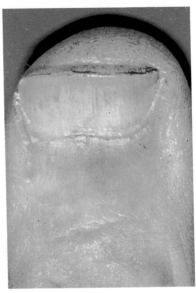

83

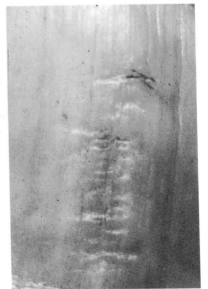

84

82 Damage from nail biting. Close-up of fingernail of an anxious clerk.

83 Personality difficulties? Nail biting – fingers from a 55-year-old, rather simple driver who lost his job through heavy drinking. Note short nails, lack of care and pigment.

84 Habit tic. Central damage to right thumb due to life-long habit 'tic'. Nail scratched on right upper canine tooth under stress!

Chapter 2

Growth Arrest or Beau's Lines

First described by the French author Beau (1846), these transverse lines in the nails are an invaluable sign of previous illness. To be due to significant illness, these lines need to be seen on all nails in one or preferably both hands. Although no excellent documentation exists, common observation indicates that a serious febrile illness for several days, or a less serious metabolic upset for a week or so is required to produce such lines. Sudden, very severe, short episodes of illness such as life-threatening haemorrhage with hypotension, can produce such discernible transverse ridges across all nails.

The value of the observation lies in a standard examination of the nail in all subjects. Knowing that the growth rate from base to free edge is about 3 months in a young adult and about 4 months in an elderly person assists in dating one or more episodes of illness and concurrent growth arrest. It should be noted that in chronic ill health, arterial insufficiency or inadequate states of nutrition, the growth rate of the nail may be slower. This occurs especially in the toenails of elderly men with arterial insufficiency. Here metabolic or growth arrest lines can take up to 8 or 12 months to grow out to the free edge.

It has also been suggested that zinc deficiency may produce lines of Beau. Other studies suggest that the use of polarised light may be helpful in examining fingernail ridges. In order to see such growth arrest lines easily, the fingernails must be held at the correct angle to the oncoming light. The transverse ridges will then be highlighted across the normal shine of the nails.

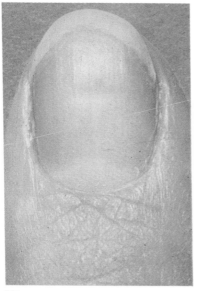

85

86

85 Single Beau's line. Beau's line in young woman with post-operative complications some 2 months previously.

86 and 87 Slightly differing views of same finger showing growth arrest line.

88 Recurrent illness. Thumbnail from 28-year-old hotel receptionist with bronchial asthma. Intermittent high steroid dosage has led to a series of growth arrest lines (Beau's lines).

89 Beau's from the side. Two severe growth arrest lines in a 34-year-old female with sub-acute bacterial endocarditis, initially partly treated. Note early 'filling' in of nail fold but no definite clubbing.

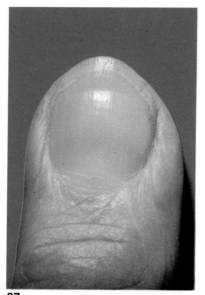

87

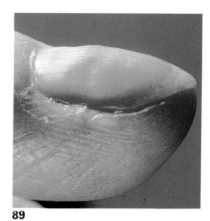

89

88

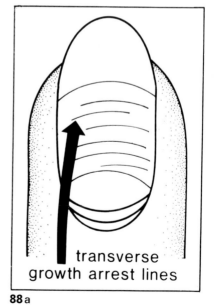

transverse
growth arrest lines

88a

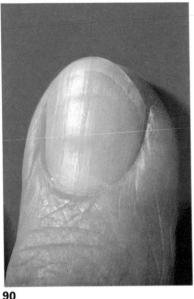
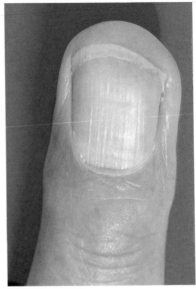

90

91

90 Evidence of two illnesses. Two growth arrest lines in nurse with severe bleeding episode and splenectomy 6 weeks later.

91 With other signs. 72-year-old hypertensive woman with serum albumen 3.2 g/l and poor nutritional habits. About 5 months previously minor myocardial infarction with resultant Beau's lines on all nails.

92 Variations in lines. Beau's lines in young secretary with repeated attacks of asthma. Note marked growth arrest line near tip of nail and minimal lines near the lunula.

93 Severe illness. Severe growth arrest for 3 months in an elderly woman who was severely ill during this period of time after post-operative and successful bowel cancer resection.

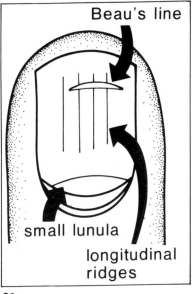

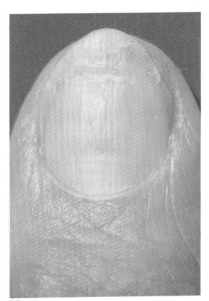

91a

92

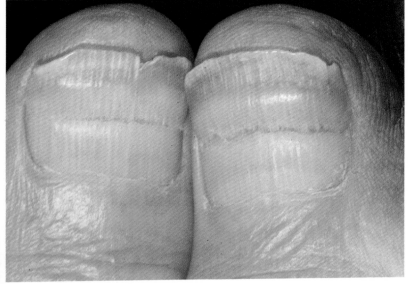

93

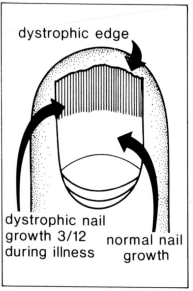

dystrophic edge

dystrophic nail
growth 3/12
during illness

normal nail
growth

93a

94

94 Complete growth arrest. Complete and severe arrest line. Major and severe life-threatening illness 6 weeks previously.

References

1. Apolinar, E. & Rowe, W. F. Examination of human fingernail ridges by means of polarized light. *Journal of Forensic Science*, 1980, 25(1), 154.
2. Beau, J. H. S. Certain caracteres de semeliologie retrospective, presentes par les ongles. *Archives in General Medicine*, 1946, 9, 447.
3. Ward, J. A. *Clinical Methods*, Hurst, J. W. (ed), Butterworth, Boston, 1981, p.542.
4. Weismann, K. Lines of Beau: possible markers of zinc deficiency. *Acta Dermato-Venereologica (Stockh)*, 1977, 57(1), 88.

Chapter 3

Koilonychia

This descriptive term is derived from the Greek word 'koilos' or spoon. In early koilonychia the nail becomes increasingly flattened, later developing a true concavity. The nail surface is often smoother than normal, especially in early koilonychia.

As there are many causes of koilonychia, the thickness of the nail plate is a reflection of the underlying disorder. Thus, koilonychia associated with iron deficiency usually shows softer and thinner nails. The thinness may however be due to the associated poor nutritional status and poor intake of sulphur-containing amino-acids. The normal keratin plate of the nails is formed by the incorporation of sulphur-containing amino-acids into the cells. To do this, amino-acid transport must be normal. Insulin and growth hormone must be present in permissive amounts to allow such transport.

Thus, patients with diabetes and less than adequate amounts of circulating insulin (i.e., 'poorly-controlled' diabetics) with mean daily blood glucose values two or three times above normal, may show flattening of the nails and later, koilonychia. In such koilonychia other nail dystrophy such as thickening may be seen.

Metallo-enzymes containing iron are also needed to form normal nail and koilonychia can be found in both iron deficiency anaemia and reduction in total body iron and low iron stores without actual anaemia.

As several biochemical steps are required to form normal nail, a genetic failure at any one of these stages may result in koilonychia. The nail patella syndrome is one such rare example.

As the returning arterioles from the digital pulp plexus enter the nail matrix from each side, it suggests that deficiencies in nail formation as well as flattening, may well be seen centrally in the first instance. This hypothesis is at least as convincing as that put forward by Stone who suggests koilonychia to be a change of angulation of the principal matrix secondary to connective tissue changes. He suggests that spooning will occur if the distal end of the matrix is relatively low compared with the proximal end.

The essential point about flattened nails and koilonychia is to ascertain whether the condition has always been present. In some families, growth hormone demand in early life may give rise to some koilonychia which should disappear and not be present after primary school age.

Koilonychia may also be secondary to occupational hazards such as solvents, to local injury, to psoriasis, or in association with Raynaud's phenomenon. If not occupational, 90 per cent of patients with koilonychia seen in general practice or internal medicine have this fingernail abnormality due to three main conditions. In the list below, conditions 1 to 3 account for almost all cases in a series reviewed by the authors from a major general teaching hospital.

95 Koilonychia. Table to show conditions which may give rise to koilonychia. Sulphur-containing amino-acids require growth hormone and insulin to be used in cells of nail bed and metallo-enzymes containing iron are rate-limiting in this step. Hence, gross protein-calorie deficiency specific dietary lack of cysteine or methionine or iron deficiency will all produce koilonychia as also will long-term insulin deficiency.

95

Koilonychia

(1) Iron deficiency
(2) Sulphur-containing amino-acids ↓
(3) ♂ Diabetics 15 to 20 years disease
(4) Raynaud's disease (uncommon)
(5) Developmental abnormality

96 Early koilonychia. 'Drop of water test': An old-fashioned but once popular test for spooning of nails. General appearance and normality of nail are more useful additional signs in early koilonychia.

97 Koilonychia as part of deficiency state. Thin flat nails in a 67-year-old woman with anaemia due to nutritional deficiency (documented low albumen, iron and folate). Note also loss of lunula and poor manicure, suggesting loss of interest in personal state.

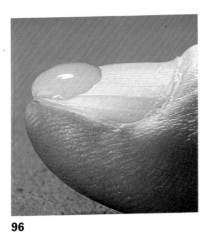

96

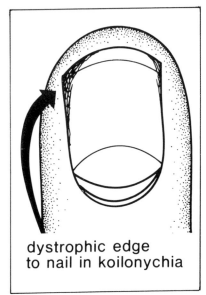

dystrophic edge
to nail in koilonychia

96a

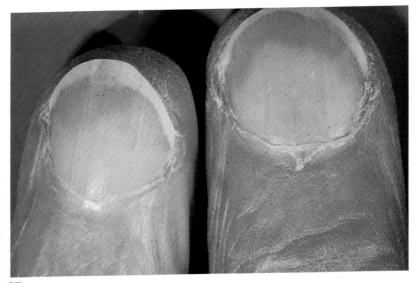

97

 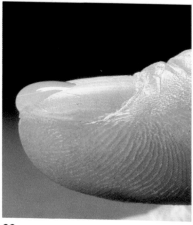

98   **99**

98 Koilonychia in diabetes. Koilonychia – diabetic nail with high blood sugars, raised growth hormone and nail dystrophy in a male aged 37.

99 Water test. Koilonychia. Drop of water test.

100 Koilonychia – nail variation. Dorsal view of hand with koilonychia – shows marked variation in degree of flattening or spooning in different nails.

101 Koilonychia and thin nails. Koilonychia – fingers from older woman with combined anaemia and nutritional deficiency.

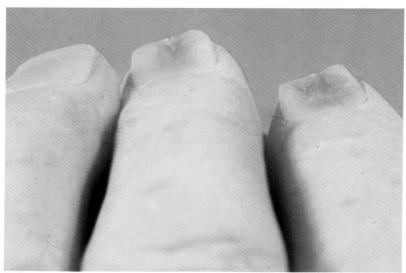

100

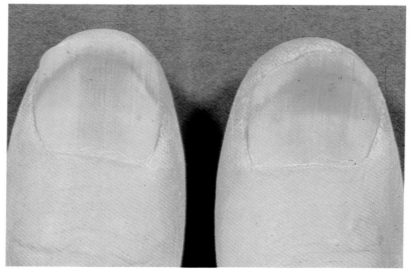

101

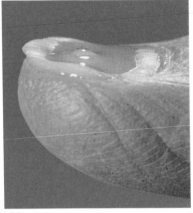
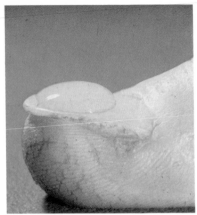

102   **103**

102 Combined deficiency. Severe spooning with some thinning of the nail in 72-year-old depressed woman with combined deficiency of iron and protein intake.

103 Free edge dystrophy. Although the thumb or great toe (1st toe) have the greatest volume of nail and thus show maximal changes, other nails also show changes.

104 Insulin deficient koilonychia. Marked nail dystrophy and koilonychia in a middle-aged male diabetic. Serum and bone marrow iron were normal but there was moderately severe insulin deficiency over 10 to 15 years.

105 Koilos in short thumbnail. Koilonychia – same 37-year-old male diabetic. Note thickness of dystrophic nail in clerical worker. ?Due to excess sugar and growth hormone.

106 Raised free edge. 'Drop of water' test for koilonychia.

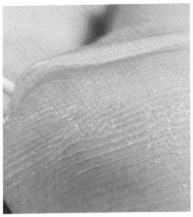

104

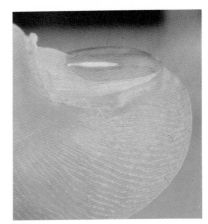

105

106

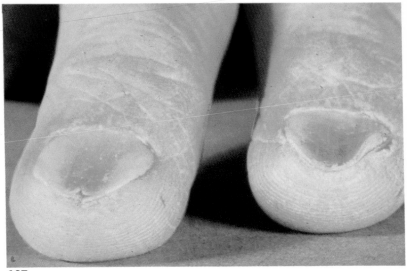

107

107 Severe classical koilonychia. Severe long-term iron deficiency resulting in gross koilonychia in a 60-year-old woman of reduced intelligence.

108 Koilonychia, pigment and nail splitting. Severe koilonychia with pigmentation and splitting of the nail's free edge.

109 Dramatic spooning. Close-up of free edge of nail to show dramatic 'spooning' and thickening in koilonychia.

110 Multiple signs in toe. Great toenails also show the koilonychia seen best in the thumb nails. Note also old injury, paronychia and poor manicure.

108

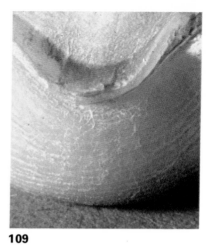

109

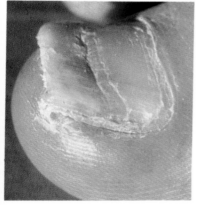

110

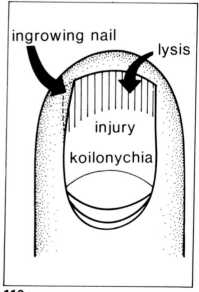

110a

111 Diabetic toe koilonychia. Severe koilonychia B in a middle-aged diabetic man with normal serum iron and no anaemia. Note damage to toe pulp.

113 Analysis of nail clippings. Cysteine content of middle-aged diabetic males with poor blood sugar control.

111

SULPHUR IN THUMB NAILS

● **Controls**

	7 papers = 3.15%	1914 – 37 (2.7 – 3.8)

● **ChCh controls**

	= 3.24%	(3.2 – 3.34)

● **Diabetics**

	= 2.9%	(2.1 – 3.2)

SIG diff .01

112

CYSTEINE IN NAILS %

- 1933   –   5 papers
- Normals – 11.1%
            (8.8 – 13.5)
- Diabetics – 8.7%
    (6)   (6.3 – 9.8)

SIG .01 level

113

References

1. Bergeron, J. R. & Stone, O. J. Koilonychia. *Archives of Dermatology*, 1967, **95**, 351.
2. Bentley-Phillips, B. & Bayles, M. A. H. Occupational koilonychia of the toe nails. *British Journal of Dermatology*, 1971, **85**, 140.
3. Beutler, E. Tissue effects of iron deficiency. In Gross, F. (ed), *Iron Metabolism*. Springer-Verlag, Berlin.
4. Dawber, R. Occupational koilonychia. *British Journal of Dermatology*, **91**, Supplement 10, 11.

Chapter 4

Canaliformis Deformities and Nutritional Abnormalities

Malnutrition of total protein and amino-acid intake, deficiency of sulphur-containing amino-acids and lack of iron have all been mentioned as causes of koilonychia. As the malnutrition becomes more severe, more marked nail changes have been reported. Thus, long-standing vitamin C and B$_2$ deficiencies are also said to cause koilonychia. Various reports incriminate a negative calcium balance and deficiency of divalent ions such as magnesium with soft or even brittle and hard nails.

In children with kwashiorkor, increased nail hardness appears to be due to abnormal increases and uneven distribution of the calcium (Robson & Brooks, 1974).

In the mid-1980s, quick clinical methods for studying individual trace mineral deficiencies giving rise to changes in the nails, have not been clearly established, although earlier work did appear promising (Harrison & Clemena, 1972; Hopps, 1977). Sometimes a specific abnormality may be detected in the hair or even toenails, but not the fingernails as shown for zinc (McKenzie, 1979). Under circumstances of clinical suspicion where specific nutrient deficiency may be present and where abnormal softness, abnormal brittleness or dystrophy is present, further clinical examination and plasma biochemistry will be necessary.

If the nail complaint is of brittleness, a standard method of springing or flicking the free nail edge, as shown, should be used. Brittle nails may occur in general systemic diseases in older patients where there is a reduced intake of the main macro- and micro-nutrients.

Brittle nails may also occur from repetitive water immersion or occasionaly in more general skin disorders such as alopecia totalis. In some rare and complex congenital syndromes, skin and nail changes may be seen with both brittleness and thickening.

In internal medicine practice, canaliformis deformities are almost always seen in association with either changes in nutrition or local central trauma to the nail. Rare familial cases have been described. This canaliformis deformity was never found in a group of healthy elderly people surveyed for nutritional abnormalities by the authors.

 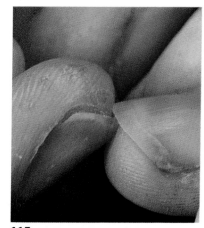

114   **115**

114 Flexibility downwards. Method of flicking upwards and downwards of the nail edge. This is to measure the softness or flexibility of the free edge. Particularly useful in nutritional deficiencies.

115 Flicking nail upwards.

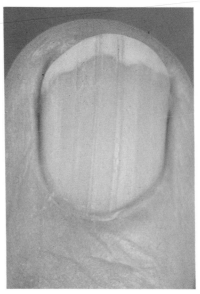

116

116 Nutritional deficiencies. Brittle thin nails in middle-aged woman on thyroxine replacement for surgical induced hypothyroidism. Note also marked longitudinal ridges and loss of lunula. Serum calcium low.

117 Early canaliformis. Very early medial canaliformis deformity in a young woman with obsessive food ideas. Little or no intake of calcium in her diet.

118 Nutritional canaliformis. 'Canaliformis' deformity in elderly anaemic woman with combined iron, folate, mineral and protein deficiency.

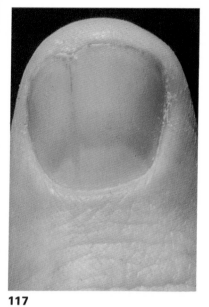

117

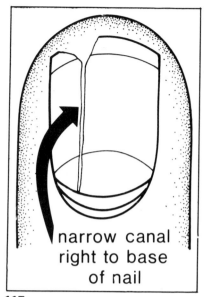

narrow canal
right to base
of nail

117a

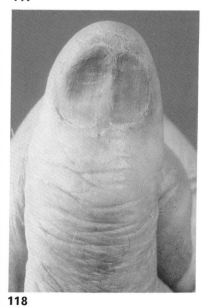

118

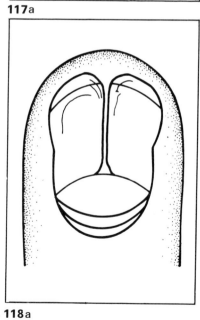

118a

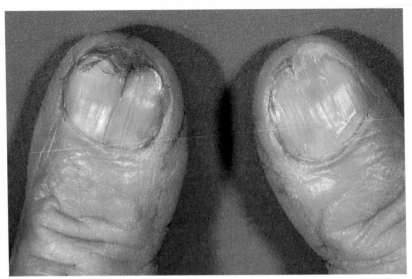

119

119 Variation in nails. Canaliformis deformity with differences between nails.

120 Distal deformities predominate. Dystrophic nails in a 67-year-old pensioner with nutritional deficiency secondary to extensive gastric surgery. Early canaliformis deformity of left index finger.

121 Chevron deformity Leclercq type canaliformis. An interesting early canaliformis deformity of the pseudo-Leclercq variety. In this young clerical worker this permanent lesion is due to very localised matrix injury some years before.

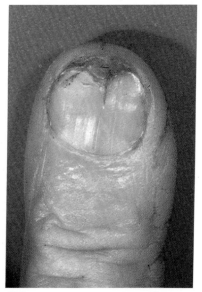

120

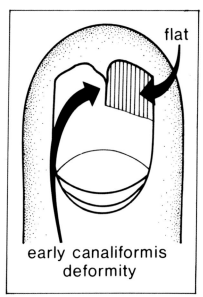

flat

early canaliformis
deformity

120 a

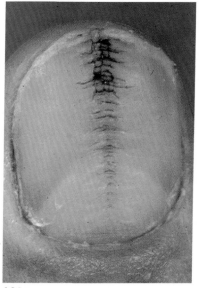

121

71

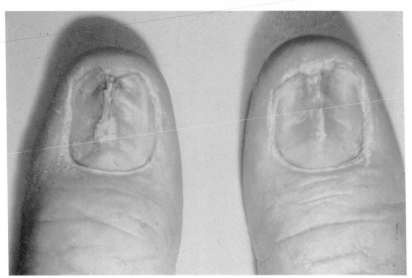

122

122 'Hellers' type. Hellers type of dystrophia mediana canaliformis in an elderly man with arterial disease.

123 and 124 Severe canaliformis. Dystrophis mediana canaliformis in a retired elderly male living alone and associated with abdominal discomfort. Mild iron deficiency anaemia and poor nutritional state. Note oblique grooves described as chevron shaped canaliformis deformity of Leclercq – can be traumatic as well as nutritional.

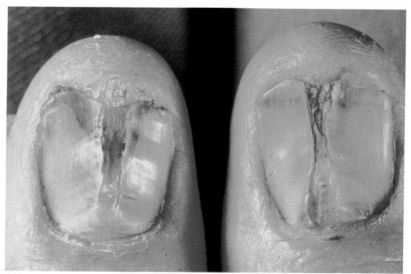

123

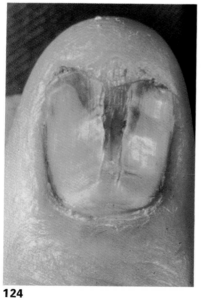

124

References

1. Abdel-Aziz, A. H. & Abdel-All, H. Dystrophia unguium mediana canaliformis. *Cutis*, 1979, **23**(3), 344.
2. Harrison, W. W. & Clemena, G. G. Survey analysis of trace elements in human fingernails by spark source mass spectrometry. *Clinica Chimica Acta*, 1972, **36**, 485.
3. Heller, J. Dystrophia unguium mediana canaliformis. *Dermat. Z.*, 1928, **51**, 416.
4. Hopps, H. C. The biologic basis for using hair and nail for analyses of trace elements. *Sci. Total Environ.*, 1977, **7**(2), 71.
5. Leclercq, R. Naevus striatus symetricus unguis, dystrophie mediane canaliforme de Heller ou dystrophie ungueale mediane en chevrons. *Bull. Soc. fr. Derm.*, 1964, **71**, 654.
6. McKenzie, J. M. Content of zinc in serum, urine, hair and toenails of New Zealand adults. *American Journal of Clinical Nutrition*, 1979, **32**(3), 570.
7. Robson, J. R. K. & Brooks, G. J. The distribution of calcium in finger-nails from healthy and malnourished children. *Clinica Chimica Acta*, 1974, **55**, 255.
8. Shelley, W. B. & Rawnsley, H. M. Aminogenic alopecia. *Lancet*, 1965, **ii**, 1327.

Chapter 5

Paronychias

Acute inflammation under the nail fold is usually due to Staphylococci. There is rapid swelling with acute inflammation and tenderness, with or without evidence of pus. A collection of pus usually damages the cells in the matrix or nail bed, leading eventually to some nail dystrophy or nail plate damage.

Chronic paronychia is much more common and is not usually associated with more general reasons such as diabetes, alcoholism or auto-immune disease for reduced immunity, as in the acute version. Low-grade redness and swelling may be present for weeks or months. This is the commonest nail complaint. Intermittent exposure to water, dirty clothes and detergents seems to be a major risk factor. Amongst women who outnumber men 5 to 1 in this area, the commonest age of presentation is 30 to 60 years and apart from housewives, barmaids, nursery and laundry workers seem particularly prone. Amongst males, it is reported from the United Kingdom that chefs, barmen, fishermen and fishmongers are more prone.

Various bacteria are involved including commonly Staphylococci and less often Streptococci, Coliform bacilli and *Pseudomonas aeruginosa*. Some authors maintain that *Candida albicans* plays a crucial role in maintaining chronicity in what would otherwise be a short, sharp infection. Stone & Mullins report that chronic paronychia starts with invasion of the epidermis on the deep surface of the proximal nail fold. When chronic over several months, the paronychia, if generalised around the base of the nail, leads to various nail dystrophies. A good review of the various hypotheses relating to paronychias by Baran (1979) stresses the complementary role of *Candida* with other bacterial infections in producing more severe and long-standing paronychias. These chronic disorders with nail base swelling and reddening of the nail fold are a particular challenge in diagnosis and treatment.

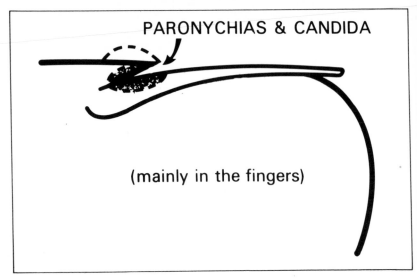

PARONYCHIAS & CANDIDA

(mainly in the fingers)

125

126 Acute bacterial infection. Acute lateral paronychia. Again in a man with brachyonychia, possibly on the basis of nail biting, but he also has previously undiagnosed diabetes mellitus.

127 Acute paronychia in great toenail. Initially suggests acute tinea but a bead of pus grew staphylococcus.

128 Herpetic whitlow. Frequently seen in nursing staff. When severe, gives onycholysis or nail shedding (La Rossa & Hamilton, 1971).

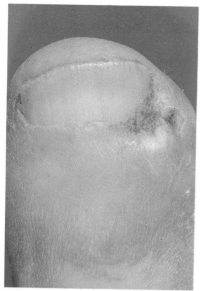

126

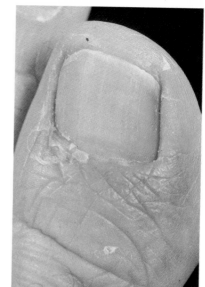

127

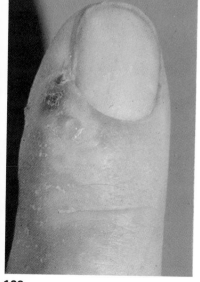

128

128 a

Herpes Simplex
Vesicle

77

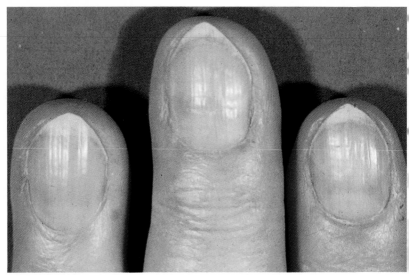

129

129 Mild healing paronychia. Healing paronychia in 23-year-old female clerical worker. Note old lesion on right side of finger.

130 Chronicity in one finger. Chronic and persistent low-grade paronychia in a woman trainee air-force chef. Note care and attention to rest of manicure.

131 Healing with dystrophy. Healed paronychia in 36-year-old married woman with six children. Consequent nail dystrophy.

132 Chronic paronychia – marked nail ridging. Chronic paronychia. Mixed infections. Note nail dystrophy and lysis on all edges. Transverse ridges on nail.

133 Secondary paronychia. Acute cellulitis of toe due to secondary bacterial infection on the basis of tinea. Note chronic nail dystrophy.

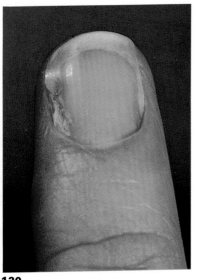

130

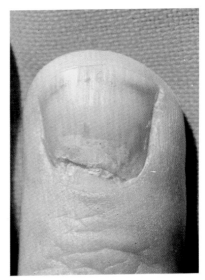

131

132

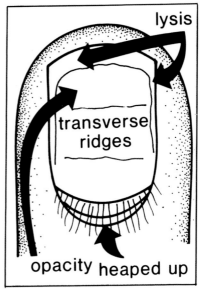

132a

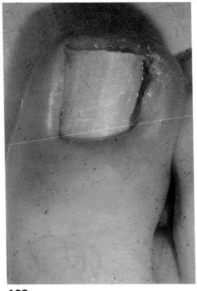
133

134

134 Repeated infections. Acute bacterial infections are common in housewives or where general immunity is impaired.

135 Small lateral lesion with pigment. Lateralised local small area of chronic paronychia leading to nail changes distally.

136 Distortion of nails. Chronic paronychia with chronic nail dystrophy in a 42-year-old housewife on long-term corticosteroid medication for asthma.

137 Chronic *Pseudomonas*. Chronic infection in great toenail bed – *pseudomonas* infection giving characteristic green colour.

References

1. Baran, R. in *The Nail*, Pierre, M. (ed). Churchill-Livingstone, London, 1981.
2. Frain-Bell, W. Chronic paronychia. *Transactions of the St John's Hospital Dermatological Society, London*, 1957, 38, 29.

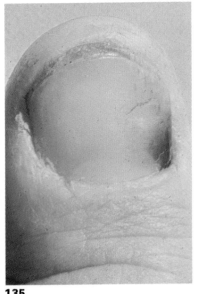

135

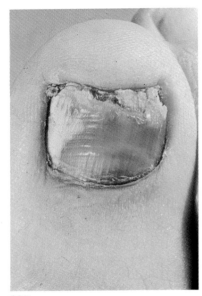

137

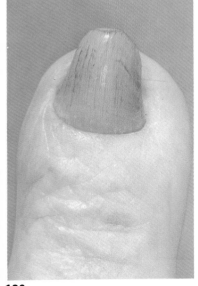

136

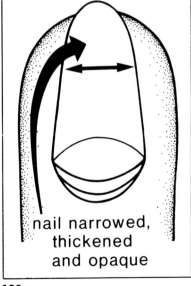

nail narrowed,
thickened
and opaque

136a

Chapter 6

Fungal Infections and Onychomycosis

Candida albicans is the commonest fungus to affect the fingernails, but in a French series of cases, 70 per cent of the involved toenail damage resulted from *Dermatophytes* and only 12 per cent from *Candida*. In the fingernails, the cause, usually *Candida*, reaches the nail plate from the lateral nail bed or works under the proximal nail fold and cuticle. The earliest sign of fungus of the fingernail will usually be small asymmetrical white patches or lifting of one edge of the nail plate. Subsequently the mycosis may spread to involve the whole nail plate with slowly increasing opacity, thickening and distortion.

As the integrity of the nail is damaged, both air and dirt may enter the split layers of the nail plate to give a variety of appearances. These include white nails, brown nails and thickened and black nails. Gradual lifting of the nail plate occurs with lateral or distal onycholysis. In more severe infections with a heavy burden of fungus, the whole nail may be shed. If the clinical appearance is not characteristic, clippings or scrapings of the nail should be taken to establish the diagnosis.

In the toes, one or occasionally more of the nails are usually involved with a gradually developing chalky opacity of the nail plate. In the toes, the most common infections are *Trichophyton rubrum* and *Trichophyton interdigitale*. Other moulds and yeasts may be found in about half the cases of fungus infection in the toe nails. The frequency of different types of infections is clearly related to the epidemiology of fungi, yeasts and moulds in that geographical area. Temperature and humidity are also critical factors.

Aspergillus, Cephalosporium and *Scopulariopsis brevicaulis* may also be found in the toenails. The latter is usually confined to one great toe, giving a characteristic chalky, yellow appearance.

As with this latter group of moulds, earlier damage to the nail or nail folds is needed for secondary fungal invasion to occur.

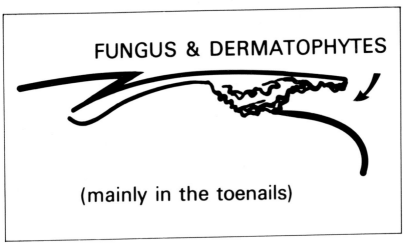

FUNGUS & DERMATOPHYTES

(mainly in the toenails)

138

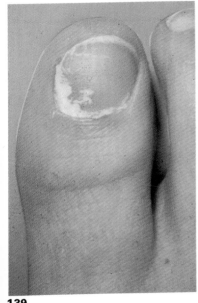

139

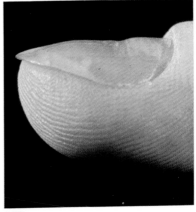

140

139 'Tinea'. Onychomycosis at base of great toenail.

140 Chronic *Trichophyton*. Lateral view of chronic trichophyton infection in index finger of young housewife with two small babies.

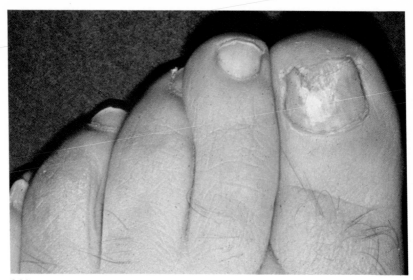

141

141 Yellow great toenail scopulariopsis. Tinea of toenail. *Scopulariopsis brevicaulis*. Note the second toe is affected.

142 Chronic fungus finger. Tinea of the fingernails in a hotel washing-up maid.

143 Chronic toenail infection. Tinea pedis. Note dystrophy mainly at distal or leading edge.

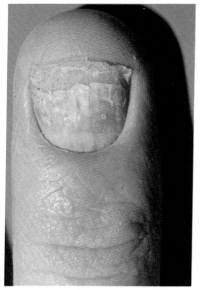

142

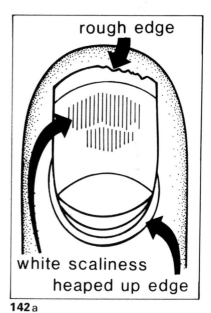

rough edge

white scaliness
heaped up edge

142a

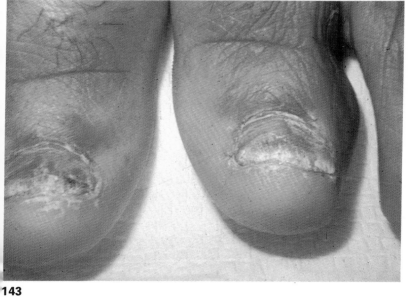

143

144

144 End-stage dystrophy. Severe end-stage nail secondary to *Trichophyton rubum*.

145 Reduced immunity. Elderly male with chronic candidiasis now under control but leading to severe nail dystrophy. Always suspect diabetes mellitus and endocrine diseases.

146 Continuing infection. Long-term chronic fungal infection *(Trichophyton)* in toenail leading to general nail dystrophy.

147 Proximal nail not infected. Dermatophytic onychomycosis. Note the disorder starts at leading edge and works back up the nail.

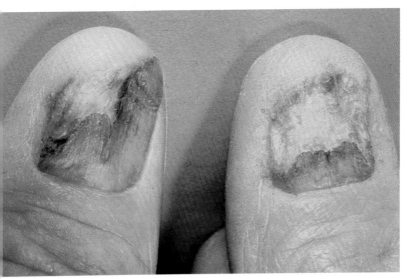

145

146

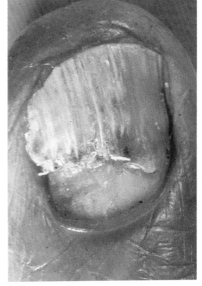

147

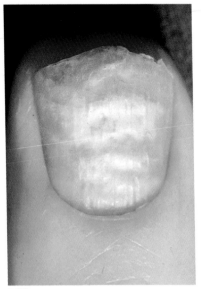

148

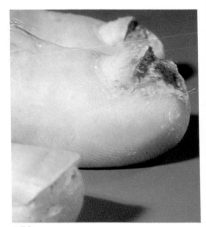

150

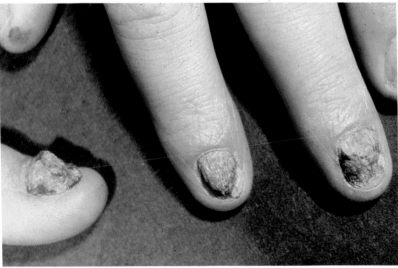

149

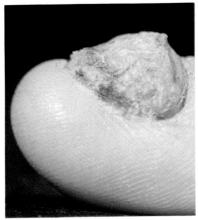

151

148 Leukonychia onychomycosis. More severe form of fungal infection in nail. Long-standing disease leads to complete opacity.

149 Nail marker of endocrinopathy. Chronic candidiasis in 19-year-old male with autoimmune endocrinopathy (including adrenal and thyroid deficiency).

150 Familial nail involvement. Very severe *Candida* infection in one of two brothers with autoimmune endocrine disease (adrenals and thyroid failure) – severe *Candida* infections may be an important marker of endocrine hypofunction.

151 Hypertrophy secondary to chronic *Candida*.

References

1. Davies, R. R., Everall, J. D. & Hamilton, E. Mycological and clinical evaluation of griseofulvin for chronic onychomycosis. *British Medical Journal*, 1967, **2**, 464.
2. Ramesh, V., Reddy, B. S. & Singh, R. Onychomycosis. *International Journal of Dermatology*, 1983, **22**(3), 148.
3. Walshe, M. M. & English, M. P. Fungous diseases of Britain. *British Journal of Dermatology*, 1966, **78**, 198.

Chapter 7

Altered Circulation

Arterial disease with atheroma in the aortic arch and the major leg vessels is common in over 65-year-old groups in Western societies. Thus, toenail changes due to atheroma and peripheral vascular disease is far more common in men than women, it occurs increasingly in later decades and is associated with cigarette smoking. In our hospital services, 95 per cent of 600 men undergoing angiography for symptoms and signs of peripheral vascular disease, had been or were heavy smokers. In contrast to this male preponderance for ischaemic changes in the toenails, women more commonly present with circulatory changes in the hands.

Raynaud's disease or episodic vascular spasm, precipitated by cold, may come on slowly over many years. It may affect mainly peripheral small vessels and thus give nail changes with predominantly less obvious generalised skin changes. Angiography of the hand is normal in Raynaud's disease and atheroma rarely, if ever, affects the hands in the first instance.

Changes in the nails with thinning and brittleness or a ground glass opacity may occur secondary to atheroma. In every such patient seen by the author, general physical examination revealed other evidence of atheroma. Signs of leg ischaemia, electrocardiographic changes, bruits over various arteries and kinked carotid arteries were always present. The only exception to this would be in cases where arm arterial emboli broke up and were not removed surgically. Under these circumstances, one sees localised and chronic changes in digital blood flow secondary to narrowing in the ulnar artery or palmar arch. Such patients are seen rarely even in a busy teaching hospital.

Raynaud's disease may be a symptomatic indication of underlying or recently developed autoimmune disease. When abnormal fingernail changes due to Raynaud's disease developing in the previous twelve months or so are seen, such patients should be fully investigated for auto-antibodies, raised sedimentation rate and other relevant testing procedures.

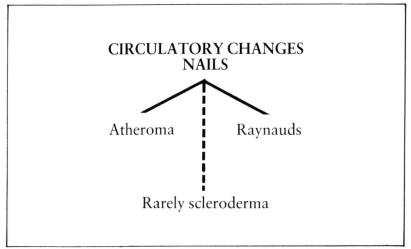

152

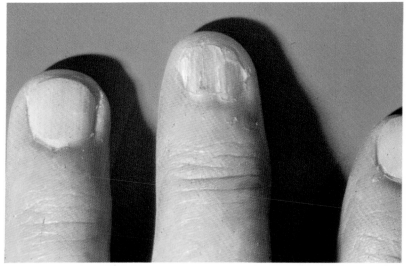

153

153 Digital artery only. Nail dystrophy secondary to injury to right digital artery to middle finger in a young male.

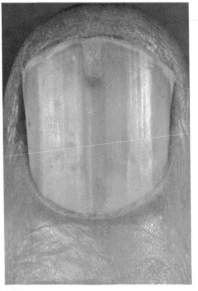

154　　　　　　　　　　　　　**155**

154 Generalised atheroma. Left thumbnail from a 69-year-old diabetic man who has severe generalised atheroma. Bilateral sympathectomy to feet; left hand often cold. Note small central onycholysis in leading edge – reddening of the centre of the nail extending through the lunula and the small and abnormal pits to the left of the mid-line. No evidence of psoriasis.

155 Severe arch atheroma. Gross aortic arch atheroma with dystrophic, flattened and opaque nail.

156 Generalised atheroma. Thumbnails of 76-year-old woman in failing health secondary to coronary artery disease and chronic left ventricular failure. Note slight cyanosis, opacity of nail, marked beading and loss of lunula.

157 Leg ischaemia. Arterial insufficiency disguised by onychogryphosis. Note ischaemic changes in skin of great toe.

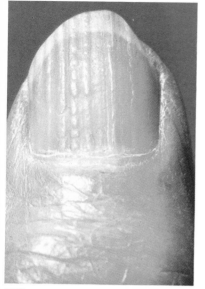

156

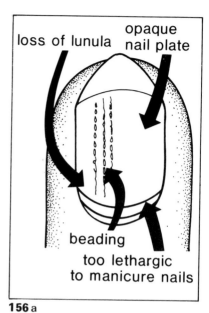

loss of lunula

opaque
nail plate

beading

too lethargic
to manicure nails

156a

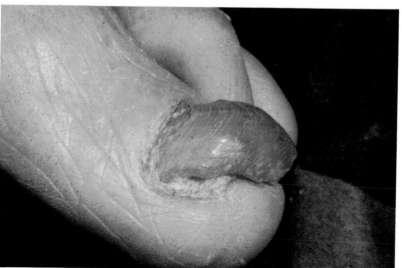

157

93

158

158 Impaired circulation.

159 Ischaemia foot. A 65-year-old man with ischaemic skin changes typical of severe reduction in blood supply. Note also scaling from dry beri-beri with abnormal TPP level.

160 Toenail ischaemia. Same 65-year-old man with absent foot pulses. Gross ischaemic changes in great toe.

162 Onycholysis in severe Raynaud's disorder.

159

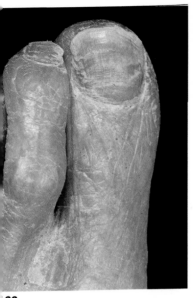

160

RAYNAUDS NAILS

- LONG RIDGES
- BRITTLE
- ONYCHOLYSIS
- PTERYGIA
- GRADUAL LOSS

161

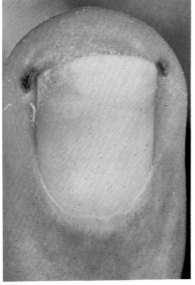

162

95

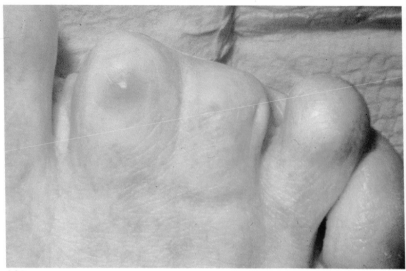

163

163 Lifelong Raynaud's. Severe Raynaud's disease with loss of toenails.

164 Early scleroderma. Nail in early scleroderma. Only change is some pointing in terminal pulp and opacity of nail.

165 Scleroderma. Some nails show pigment. Hand colour very dusky and pigmented due to secondary circulation impairment.

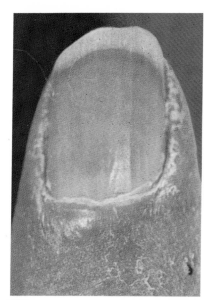

164

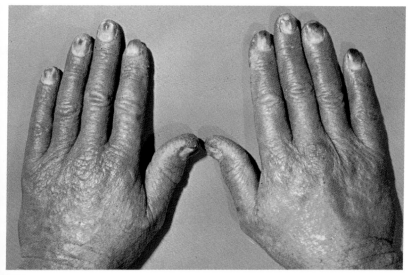

165

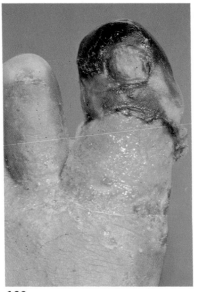
166

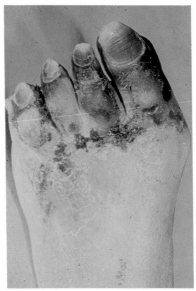
167

166 Temporary nail damage. Early frostbite – only nail of great toe will be lost. As matrix cells are damaged permanent nail dystrophy remains.

167 More marked frostbite with damage to matrix cells and toenail beds of all digits.

168 Autoimmune disease. Digital artery thrombosis in an old lady with rheumatoid arthritis and positive LE cells, leading to gangrene of fingertip and loss of nail.

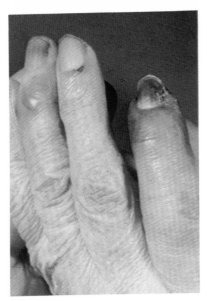

168

References

1. Edwards, E. A. Nail changes in functional and organic arterial disease. *New England Journal of Medicine*, 1948, **239**, 362.
2. Samman, P. D. & Strickland, B. Abnormalities of the finger nails associated with impaired peripheral blood supply. *British Journal of Dermatology*, 1962, **74**, 163.
3. Strickland, B. & Urquhart, W. Digital arteriography, with reference to nail dystrophy. *British Journal of Radiology*, 1963, **36**, 465.

Chapter 8

Nail Trauma

Single sudden injuries to the nail plate result in fracture of the underlying capillaries. If these are large enough and the blow severe enough, a haemorrhage under the nail develops. Very small local injuries under the dorsal nail fold may grow out. Very minor injuries may not be associated with any collection of blood and may only result in a subsequent localised whitening of that part of the nail. This, in turn, grows outward.

More severe injuries may give rise to contusion and tissue damage with actual avulsion of the nail plate. Where the haematoma is spreading under the nail plate and lifting this off the bed, a hole is usually drilled through the nail plate to reduce the pressure in the haematoma. This time-honoured procedure has not been subjected to any reported controlled trial but appears to be based on sound mechanical principles.

The management of crush injuries to the nail alone or of the fingers calls for specialised skills, preferably in a hand unit, and is well discussed by Michon & Delagoutte (1980) and by Iselin (1980).

Localised injury to the nail base sufficiently severe to produce long-term damage and pterygium formation may give rise to prominent longitudinal ridging arising from that point, or even in some patients the distinctive canaliformis deformity. Chronic trauma of the nails may arise from unusual occupational use but is more often secondary to personality or habit traits.

Thus nail-biting in adults is usually a failure to resolve personality conflicts and a lack of feeling of self-worth late in life. Habit tics of various types will damage one single nail only. Excessive anxiety or even a habit trait may lead to repeated pushing back of the cuticle with manicure equipment, 'orange sticks' or even the teeth. Such repetitive minor trauma can lead to transverse lines, nail opacities or splitting of the free edge of the nail.

'Hang nails' are torn or narrow strips of the horny epidermis split up from the lateral nail fold. Failure to care for the cuticle, nail-biting or using the teeth to push back the quick may lead to an increased frequency of damage.

white spot of
slight trauma
3 months previously

169

169a

169 White spot of slight trauma 3 months previously. No nail dystrophy.

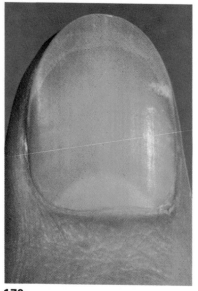 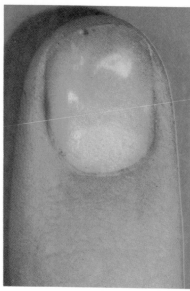

170 Mild unremembered injury. Normal nail with small old (white) injury spot on left border. Female scientist in her 40s.

171 Frequent knocking of hands. Mild trauma resulting in one large and several small white patches.

172 Occupational injury. Multiple small injuries over 3 or 4 months in a young woman working at basket-making. Many small white spots whereas only a few are seen in 'normal' nails.

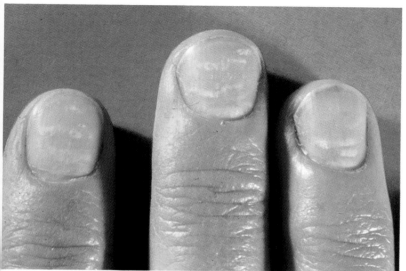

172

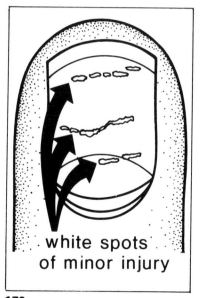

white spots
of minor injury

172a

103

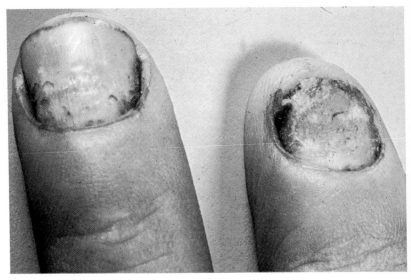

173

173 Occupation with trauma. Dystrophy and minor repeated chemical injury to eponychium or nail fold in a factory worker in car battery trade.

174 Car door injury. Fingernail from senior nursing sister whose finger was lightly jammed in a car door. Partially disguised by nail varnish, can be a growth arrest line.

175 Subungual haematoma. Thumbnail of middle-aged male sustained moderate trauma from hammer injury weeks previously. Note hole drilled in attempt to release haematoma.

176 Healing subungual haematoma. Healing phase of 175 photograph taken 8 weeks later showing somewhat increased rate of growth of new nail plate.

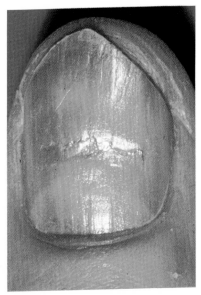

174

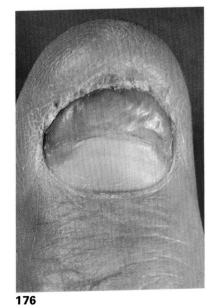

176

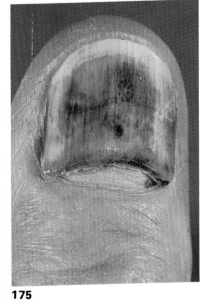

175

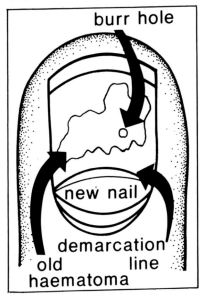

burr hole

new nail

demarcation
old   line
haematoma

175a

105

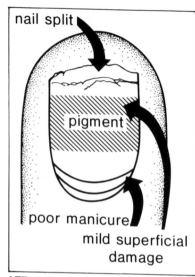

177

177a

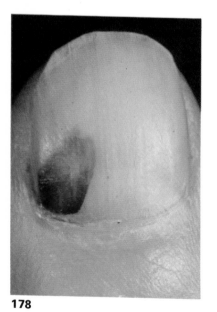

177 Major occupational trauma. Nail dystrophy due to repeated work trauma in printing trade.

178 Repeated nail bleeding. Mild trauma leading to subungual haematoma in a long-standing diabetic. Positive hess or cuff test on arm and normal platelets suggests capillary fragility.

178

179 Childhood trauma. Trauma to middle fingertip, including nail matrix from an injury in early childhood. Could be reconstructed by some hand units. Nail grafting in a female could be undertaken.

180 Dorsal view. Moderate trauma to nail some years previously. Pigment probably due to low grade inflammatory bowel disease.

181 Lateral view of old injury. Moderate trauma to nail some years previously. Pigment probably due to low grade inflammatory bowel disease.

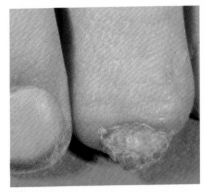

179

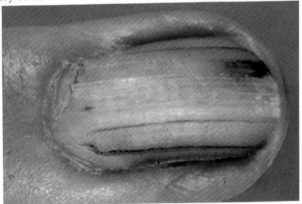

180

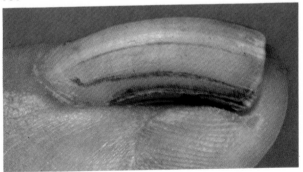

181

182

183

182 Fingertip trauma. Small left-sided pterygium in the fingernail of a labouring man who suffered a stroke from severe arterial disease.

183 Repetitive trauma. Nail dystrophy due to repetitive work trauma in wire worker.

References

1. Farrington, G. H. Subungual haematoma – An evaluation of treatment. *British Medical Journal*, 1964, **1**, 742.
2. Johnson, R. K. Nailplasty. *Plastic and Reconstructive Surgery*, 1971, **47**, 275.
3. Samman, P. D. A traumatic nail dystrophy produced by a habit tic. *Archives of Dermatology*, 1963, **88**, 895.
4. Samman, P. D. Nail disorders caused by external influences. *Journal of the Society of Cosmetic Chemistry*, 1977, **28**, 351.
5. Stone, O. J. & Mullins, J. F. The distal course of nail matrix haemorrhage. *Archives of Dermatology*, 1963, **88**, 186.
6. Michon, J. & Delagoutte, J. P. (p.81) & Iselin, M. (p.83) In *The Nail*, Pierre, M. (ed). Churchill Livingstone, London, 1981.

Chapter 9

Other Onychodystrophies

Lifting of the nail plate from the nail bed is literally a lysis of the nail – onycholysis. The commonest causes in general medical patients are thyrotoxicosis, trauma, psoriasis and elongated nails with or without secondary infection. Other rare congenital, drug-induced and occupational onycholyses also occur. Onycholysis is also reported in lichen planus and the specific disorder of lichen striatus where there are papules on the skin of the affected limb seen in association with onychodystrophy.

In a review of 113 cases of idiopathic onycholysis of the great toenail, Baran & Badillet (1982) found trauma to be a major aetiological factor. In many of their cases, trauma was found to be associated with the wearing of inappropriate shoes.

In other skin diseases associated with onycholysis, the cause of the nail lifting and shedding is usually obvious as in contact dermatitis and chemical and solvent injuries.

Common onycholysis:	Trauma
	Thyrotoxicosis
	Psoriasis
	Long nails and trauma
Cutaneous (less common):	Dermatitis (often contact)
	Congenital nail syndromes
	Drug eruptions (Bleomycin, Doxorubicin)
	Side effects of treatment and photosensitivity
	to drugs (Tetracyclines, Chlorpromazine)
Local and chemical causes:	Occupational
	Chemical: alkalis, solvents, cosmetics

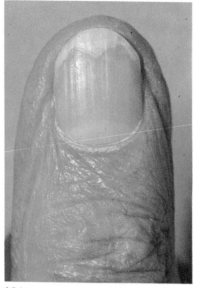
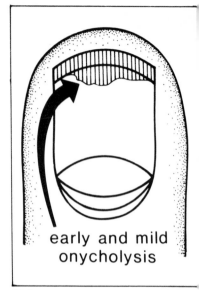

early and mild
onycholysis

184   **184a**

184 Onycholysis from anaemia. Distal onycholysis in 59-year-old woman with breast cancer and recent anaemia from secondary bone marrow involvement. Nutrition and serum albumen normal. Note: there is no koilonychia.

185 Lateral onycholysis. Three months history of lateral onycholysis in young male dentist. ?Hand trauma.

186 Multiple changes. Nail changes with lateral onycholysis, thinning of nail plate and growth arrest lines in a 72-year-old diabetic with long-standing requirement for insulin. Only general medical abnormalities found were indifferently controlled diabetes mellitus and very poorly fitting upper and lower dentures.

185

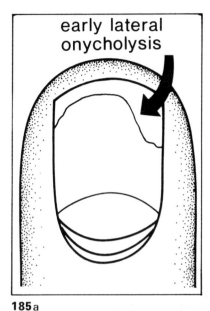

185a

early lateral
onycholysis

186

lateral onycholysis

growth arrest lines

reddening due to
thinning of nail plate

186a

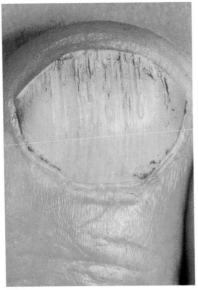
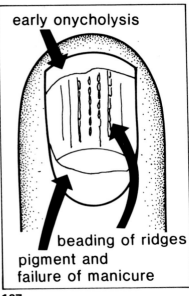

187   **187a**

187 Multiple changes. 'Beading' on finger nail of hypothyroid male on treatment with thyroxine. Note lack of beading near lunula and failure of care suggesting lapse in treatment.

188 Onycholysis of unknown cause. Unusual onycholysis with no other abnormality other than long-standing diabetes.

189 Psoriatic onycholysis. More severe onycholysis in a 34-year-old male technician due to psoriasis. Only other skin lesions were pitting in other nails and lesions in scalp hair.

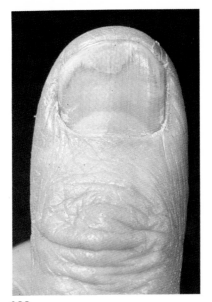

188

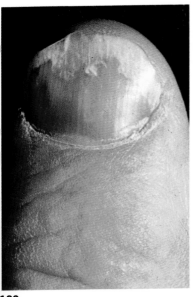

189

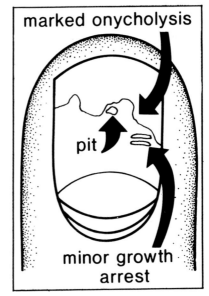

marked onycholysis

pit

minor growth
arrest

189a

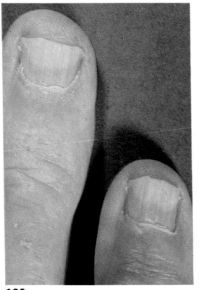

190

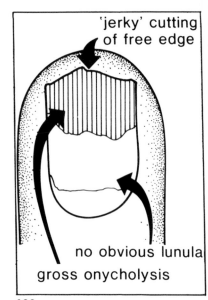

'jerky' cutting
of free edge

no obvious lunula

gross onycholysis

190 a

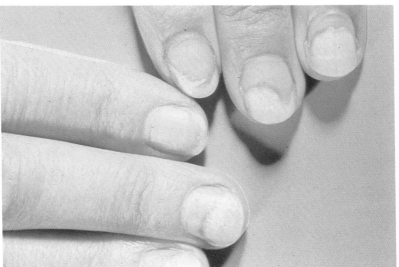

191

190 Thyrotoxic onycholysis. Untreated thyrotoxicosis in a young middle-aged woman with 1 year of increasing symptoms. No evidence of psoriasis. Note 'jerky' cutting of free edge.

191 Severe unexplained changes. Onycholysis of moderately severe degree in middle-aged woman. No known associated general or skin disease. 'Idiopathic' onycholysis.

192 Half and half onycholysis. Unexplained lysis of most nails in middle-aged woman. No evidence of psoriasis but needs follow-up.

193 Widespread non-psoriatic onycholysis – drug induced. Drug-induced by tetracylines and chlorpromazine in a woman in her 60s.

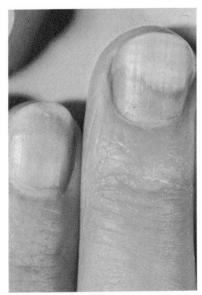

192

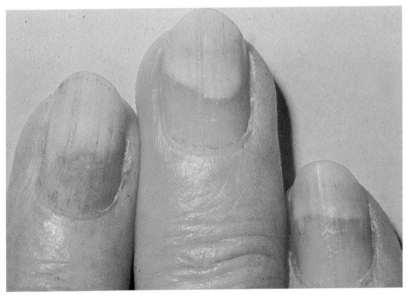

193

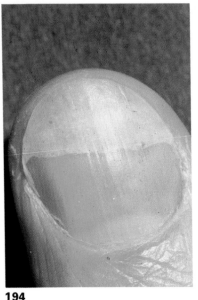

194

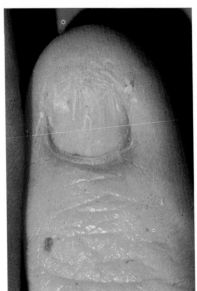

195

194 Severe thyrotoxic onycholysis. 'Half and half' onycholysis seen in middle-aged woman with severe thyrotoxicosis.

195 Unknown dystrophy. Unknown terminal nail dystrophy. No known psoriasis or other skin condition. This will be watched in the next few years.

196 Ventral spur only. Central ventral spur on underside of nail only – not associated with any known disease.

197 Nail dystrophy. Microcryptosis in smallest toenail.

198 Lichen planus. Distal nail dystrophy in young female with lichen planus.

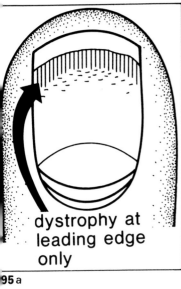

dystrophy at
leading edge
only

195a

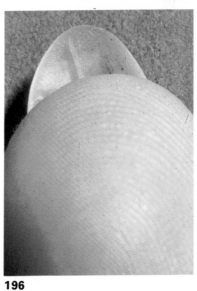

196

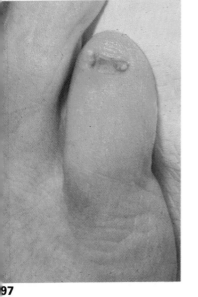

197

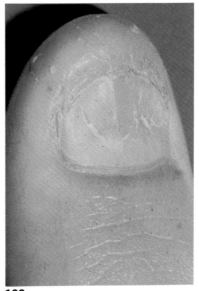

198

117

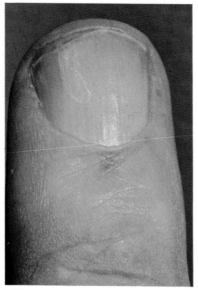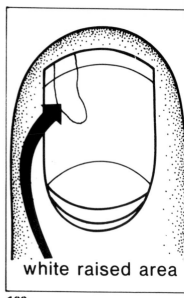

199 **199a**

199 Early lichen planus. Very early lichen planus giving rise to odd raised whi
area distally.

References

1. Baran, R. & Badillet, G. Primary onycholysis of toenails (113 cases). *British Journal of Dermatology*, 1982, **106**(5), 529.
2. Baran, R. & Temime, P. Les Onychodystrophies Toxicomedicamenteuses e les Onycholyses. *Concours medicine*, 1973, **95**, 1007.
3. Dawber, R. P. R., Samman, P. D. & Bottoms, E. Finger nail growth in idiopathic and psoriatic onycholysis. *British Jounal of Dermatology*, 1971, **85**, 558.
4. Eastwood, J. B., Curtis, J. R., Smith, E. K. M. & Wardener, H. E. de. Shedding of nails apparently induced by the administration of large amoun of cephaloridine and cloxacillin in two anephric patients. *British Journal of Dermatology*, 1969, **81**, 750.
5. Frank, S. B., Coher, H. J. & Minkin, W. Photo-onycholysis due to tetracycline hydrochloride and doxycycline. *Archives of Dermatology*, 197 **103**, 520.

6. McCormack, L. S., Elgart, M. L. & Turner, M. L. Benoxaprofen-induced photo-onycholysis. *Journal of the American Academy of Dermatology,* 1982, 7(5), 678.
7. Myers, M., Storino, W. & Barsky, S. Lichen striatus with nail dystrophy. *Archives of Dermatology,* 1978, **114**(6), 964.
8. Orentrich, N., Harber, L. C. & Tromovitch, T. A. Photosensitivity and photo-onycholysis due to demethylchlortetracycline. *Archives of Dermatology,* 1961, 83, 730.
9. Ray, L. Onycholysis. *Archives of Dermatology,* 1963, 88, 181.
10. Swan, R. H. Oral lichen planus with associated nail changes. *Journal of Oral Medicine,* 1982, 37(1), 23.
11. Wilson, J. W. Paronychia and onycholysis. Etiology and therapy. *Archives of Dermatology,* 1965, **92**, 726.

Chapter 10

Psoriasis

'Pitting of the fingernails with lifting and flaring is supportive of the diagnosis of psoriasis and a change suggestive of an oil droplet underneath the nail is pathognomonic.' This quotation from Cecil's *Textbook of Medicine* (1982) emphasises the importance of fingernail examination in psoriasis. This disorder may begin at any age but more usually in young adults. The peak incidence is in people in their 30s. HLA tissue typing markers are useful only in determining whether there is some inherited component. Patients with a tissue type of HL-A$_{13}$ have milder disease and little heritable tendency, where W$_{17}$ patients have a significant number of affected relatives and an earlier age of onset.

Psoriasis is without a specific known cure but remissions which were previously seasonal can be enhanced by using combinations of various tar preparations and ultraviolet light. The fingernails may be affected before there are widespread, generalised lesions although small lesions can be found in the region of elbows or knees in 95 per cent of patients with definite psoriatic changes in the nails. In a few unusual cases, the nail changes are said to ante-date the more generalised skin changes by up to several years. Well-documented cases are rare.

It is also well recognised that some of the worst deformities of the nails may have only minor skin involvement around the hands. Histology of the psoriatic nail may show, in increasing order of severity the following.

(1) Hypertrophy of the nail bed which may be localised or generalised. This, in turn, can give rise to thickening and distortion of the nail plate.

(2) Retention of the nuclei or parakeratosis of the nail plate. This may give rise to weakened or mottled areas and sometimes loss of lunula with otherwise relatively normal nail plates.

(3) 'Pitting' of the nail plate. These pits appear to arise from an area of softening with groups of parakeratotic cells. These latter cells drop out, leaving excavations or pits sometimes seen as punctate erosions.

(4) Onycholysis, which may be quite localised to the distal end or one side of the nail, or more general. A minor area of lateral onycholysis with elsewhere a few pits may be the only nail signs of psoriasis.

(5) Shedding of nails. In the healthy epidermis, cells starting from the basal layer take 3 to 4 weeks to reach the surface and develop into squamous cells. In psoriasis these same cells reach the surface in 3 to 4 days. Because of this seven- or eight-fold acceleration in the cell

migration, normal keratinisation does not have time to take place evenly, especially if the psoriasis is patchy. Thus rapid progress of cells reaching the surface in psoriatic zones mean that cells still have their nucleus and become parakeratotic. Sometimes the shedding is not complete but patchy. This may allow infection and microabscesses filled with polymorpholeucocytes are seen, the so-called pustular psoriatic nails.

Thus, many degrees of nail changes are found, even in the same patient, from lateral onycholysis to distal onycholysis or even whole nail shedding. Other changes seen and also reported are markedly increased thickening, over-curvature of the nail and secondary infection by fungi.

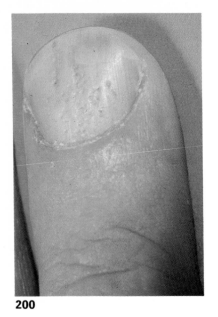
200

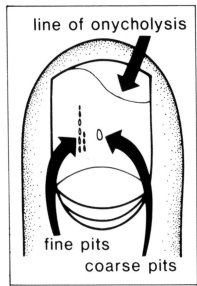
200a

200 Mild psoriasis. Mild psoriatic changes in a young-middle-aged woman.

201 Central nail plate involved. Diffuse psoriatic skin condition with diffuse but mild dystrophia shown on most of the nails.

202 Onycholysis antedating skin lesion. Early distal onycholysis led to consultation with many doctors. Eventually the small lesion on the elbow (L) led to the correct diagnosis of psoriatic nail disease.

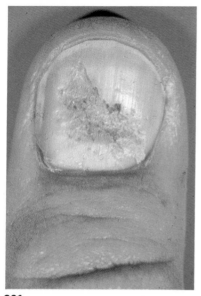

201

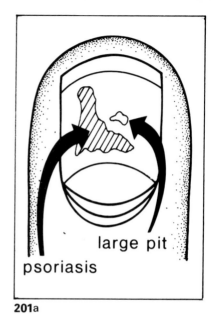

201a

large pit

psoriasis

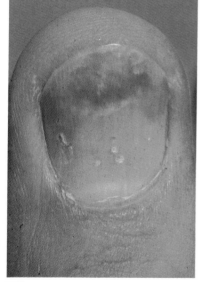

202

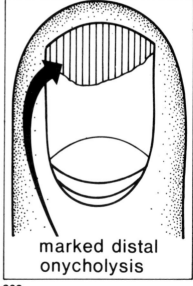

202a

marked distal
onycholysis

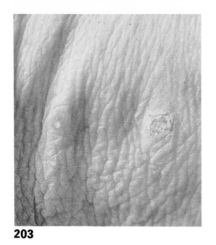

203

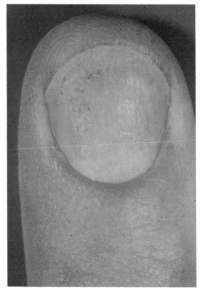

204

203 Elbow lesion of psoriasis. This was the first lesion of psoriasis to develop 2 years after onycholysis seen in 202.

204 Minimal skin but all nails involved. Opacity and flaking of all nail surface in psoriasis.

205 Elbow-knee syndrome in psoriasis.

206 Hand with nail involvement. Elbow-knee syndrome in psoriasis.

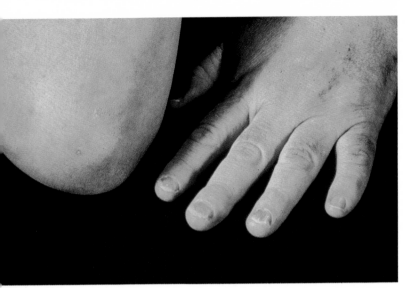

205

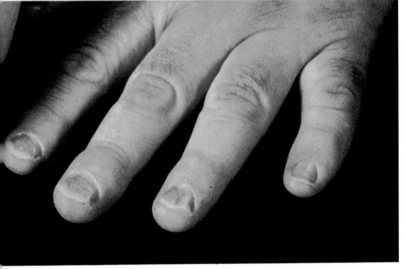

206

125

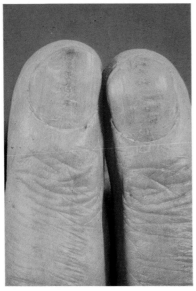
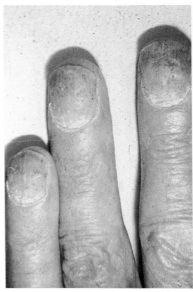

207   **208**

207 Differential diagnosis. Nail dystrophy in middle-aged woman. This is secondary to a generalised dermatitis thought to be drug-induced and now healed. Psoriasis questioned?

208 Marked psoriasis. Moderate psoriasis showing lifting of leading edge, pits, opacity, and mild generalised nail plate dystrophy. Note yellowing of nails at distal edge.

209 'Pits' only. Marked psoriatic pitting with no other skin lesions.

210 Advanced psoriasis. Gross nail dystrophy in great toe due to severe psoriasis.

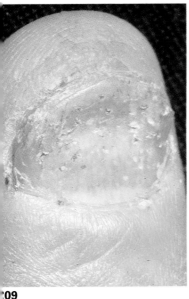

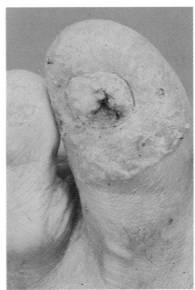

09 210

References

1. Dawber, R. Fingernail growth in normal and psoriatic subjects. *British Journal of Dermatology*, 1970, **82**, 454.
2. Dawber, R., Samman, P. D. & Bottoms, E. Fingernail growth in idiopathic and psoriatic onycholysis. *British Journal of Dermatology*, 1971, **85**, 558.
3. Lewin, K., Dewit, S. & Ferrington, R. A. Pathology of the finger nail in psoriasis. *British Journal of Dermatology*, 1972, **86**, 555.
4. Leyden, J. J. Exfoliative cytology in the diagnosis of psoriasis of the nails. *Cutis*, 1972, **10**, 701.
5. Rothenberg, S., Crounse, R. G. & Lee, J. L. Glycine-C14 incorporation into the proteins of normal stratum corneum and the abnormal stratum corneum of psoriasis. *Journal of Investigative Dermatology and Syph*, 1961, 37, 497.
6. Zaias, N. Psoriasis of the nail. *Archives of Dermatology*, 1969, **99**, 567.

Chapter 11

Clubbing of Fingers and Fingernails

Hippocrates first described fingernail clubbing in patients with empyema. One of the best and most widely known signs of general medical disease to be seen in the fingernails is marked clubbing which is easily diagnosed. The approach in this section is that of the general practitioner or general internist. Here the earliest but definite signs of clubbing are the most valuable. A proposed method of rocking the nail bed has been extensively and reliably used by one of the authors over a period of two decades. A form of grading is also proposed which is useful in practice. Equally easily, the terms 'slight', 'definite', 'moderate' and 'severe' can be used to describe increasing degrees of clubbing.

What causes clubbing? Editorials in the *Lancet* and *British Medical Journal* testify to its importance and yet the exact mechanism by which excess growth of the matrix cells is stimulated is not fully known. The best current hypothesis to fit in with known facts is that there is a combination of autonomic nerve control of the digital arterioles, together with changes in the ratio of circulating vaso-active kinins perfusing the finger pulp. Some of the blood supply carrying nutrients to the proximal nail bed and matrix passes through the very active fingertip plexus in the terminal pulp. Well recognised 'capillary shunts' here are opened and shut according to the need to lose or conserve heat. Thus, in pulmonary causes for clubbing such as dilatation of the bronchi, autonomic stimulus will play a major role, whereas in cyanotic heart disease or liver disease, changes in the vaso-active peptides could deviate more stimulated blood to the nail bed.

A common list of causes of 'simple' clubbing is set out to emphasise the value of observing early clubbing in systemic disease. Other more specific types of clubbing often given a separate classification are hypertrophic osteoarthropathy and pachydermoperiostitis. The former, usually seen in rapidly growing cancer of the lung or bronchus is really pseudo-inflammatory but is also associated with pseudoarthropathies. The latter is a rarer version of hypertrophic osteoarthropathy where a line of periosteal new bone is lifted up and can be seen on X-rays of the tips of the fingers or toes. A further sub-classification of clubbing is sometimes called the 'shell-nail' syndrome in which some long-standing bronchiectatic patients develop thin clubbed nails with atrophy of the underlying bone and nail bed.

'Clawing' of the fingertips is characteristic of long-term smokers – particularly in women. Here the pulp becomes wasted and the nail curved over but with no filling in of the proximal nail fold and no easy rockability. As the patient gradually develops increasing degrees of lung failure due to pan-acinar emphysema, the fingernails become more and more curved. With the loss of pulp, the fingertips come to resemble talons. In the early stages this goes by the name of 'beaking' as the nail curves to resemble a bird's beak. This later progresses to give a claw-like process. These are described as 'claw nails' or the condition is called 'clawing' or 'beaking'. The loss of digital pulp or lack of pulp wasting distinguishes it from clubbing.

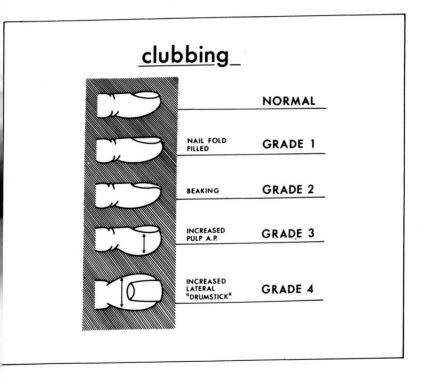

211 Grading of finger nail clubbing.

212 Method for demonstrating 'rocking'. When Grade I or early clubbing occurs, the first sign is a filling in of the nail fold with loss of angle. This area immediately over the nail fold or matrix becomes spongy. The front and back halves of the nail may be 'rocked' as in a see-saw. This may be achieved by gripping the sides of the nails between the thumb and middle finger of each hand. The two index fingers can then rock the nail up and down. The part of the seesaw is depressed with the left index finger (**212**). Then the left index finger is lifted and the right index finger depresses the most proximal end of the nail over the matrix and under the nail fold (**213**). These movements are repeated to develop a sensation of rocking with the rapidly-growing nail matrix under the nail plate as the fulcrum.

214 Measuring clubbing by X-rays. Standardised method of measuring angle and degree of clubbing from a lateral radiograph of the finger.

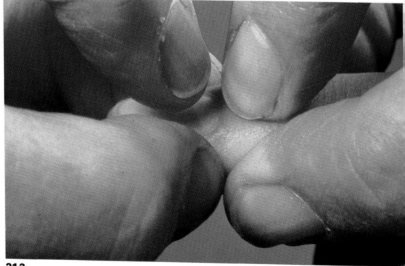

212

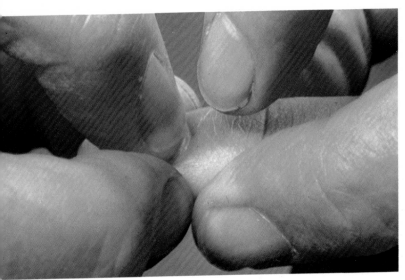

213

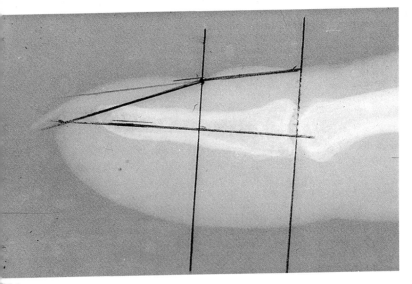

214

131

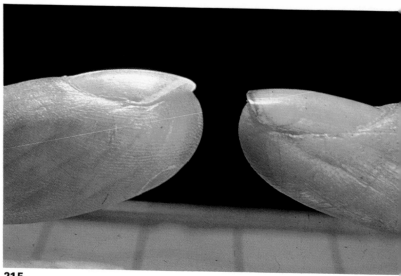

215

215 Early clubbing. Early clubbing (Grade 2) in a young woman with single lobe bronchiectasis. Comparison with age matched control (R).

216 Rapid onset clubbing. Grade 3 clubbing in a middle-aged man with fibrosing alveolitis of some duration.

217 Clubbing and other findings. Early clubbing in patient with subacute bacterial endocarditis. This man is also a smoker; note nicotine staining. Two small 'splinter' haemorrhages are also present.

218 Clubbing and pigmentation. Marked clubbing of nail in an elderly male with chronic lung disease and some bronchiectasis in a collapsed middle lobe. Also has lung failure with hypoxia. This latter appears to be the only cause of the increasing pigmentation.

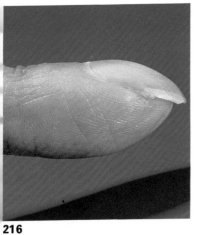

216

218

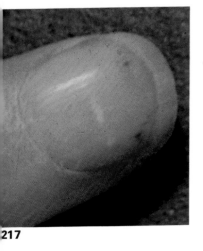

217

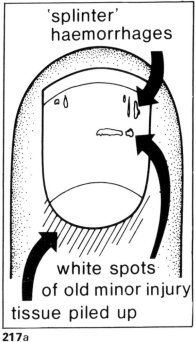

'splinter' haemorrhages

white spots of old minor injury

tissue piled up

217a

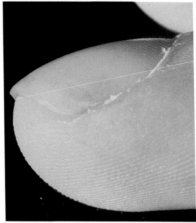

219 **220**

219 Clubbing with cyanosis. This elderly railway worker presented with cough and sputum. Note clubbing, some beaking and pulp wasting of middle finger, suggesting chronic hypoxia. Also marked cyanosis indicative of unsaturated haemoglobin, either due to polycythaemia compensatory to the longterm low oxygen levels or severe blood gas mismatch problems. Lung abscess present.

220 Clubbing in younger person. More severe clubbing in long-standing bronchiectasis. Note also anaemia and marked increase in sponginess at inner nail fold. Fluctuation of the nail bed is apparent at this site.

221 Immediate diagnosis! 72-year-old labourer who was a life-long smoker. Note strongly positive 'nicotine sign'. Presented with painful fingertips. Gross drumstick Grade 4 clubbing of nails and X-ray confirmed an inoperable bronchogenic cancer of the lung.

222 'Intestinal clubbing'. Gross clubbing (Grade 4) in a 42-year-old man who developed tropical sprue while serving in the army – with continuing diarrhoea and several bone fractures. Found to have low levels of serum calcium and phosphorus, with X-ray appearances in the bones of osteomalacia. Tests and small gut biopsy confirmed malabsorption due to small gut disease. Normal lungs, cardiovascular system and liver function.

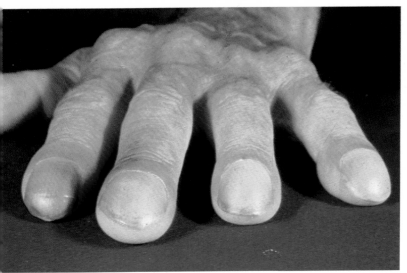

221

222

135

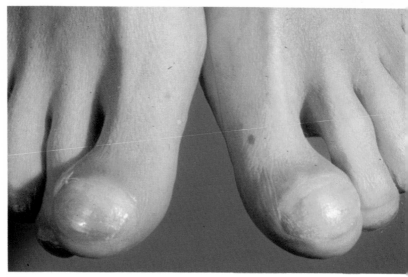

223

223 Toes are clubbed as well. Clubbing when well marked in Grades 3 or 4 is also observed in all toes as in this woman with chronic lung disease and bronchiectasis.

224 Another gut disease. Gross (Grade 4) clubbing in small gut disease.

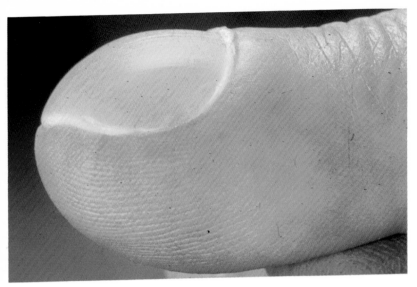

224

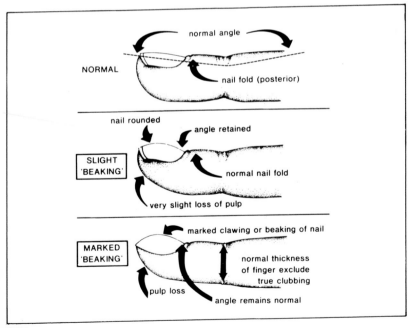

normal angle

NORMAL

nail fold (posterior)

nail rounded

angle retained

SLIGHT 'BEAKING'

normal nail fold

very slight loss of pulp

MARKED 'BEAKING'

marked clawing or beaking of nail

normal thickness of finger exclude true clubbing

pulp loss

angle remains normal

225

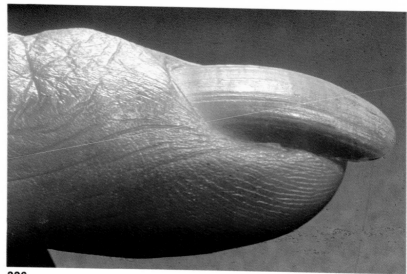

226

227

226 'Beaking' or 'clawing'. Marked nicotine staining in an elderly retired wire-worker who remained addicted to smoking with gradual loss of lung tissue due to obstructive airways disease and emphysema. Note thickening and dystrophy of nail secondary to previous occupation, beaking and pulp loss. Nicotine staining is prominent, suggesting 40 cigarettes daily. Note also Heberden's nodes. These are unusual in men and work-related.

227 'Beaking' or 'clawing'. Severe 'beaking' of nails in a heavy smoker wth hypoxia ($pAO_2 = 55$) and lung X-rays showing only emphysema with obstructive lung disease.

References

1. Leading article. Finger clubbing and hypertrophic pulmonary osteoarthropathy. *British Medical Journal*, 1977, **3**, 785.
2. The Genuine Works of Hippocrates. Translated by Adams, F. V. 849, 1: p.249, London, 1849.
3. Leading Article. Finger clubbing. *Lancet*, 1975, i, 1285.
4. Ponchon, Y., Chelloul, N. & Roujeau, J. Contribution a l'etude anatamopathologique de l' hippocratisme digital. *Semaine des Hopitaux de Paris*, 1969, **42**, 2604.
5. Racoceanu, S. M., Mendlowitz, M., Suck, A. F., Wolf, R. L. & Naftchi, N. E. *Annals of Internal Medicine*, 1971, **75**, 933.
6. Stone, O. J. & Maberry, J. D. Spoon nails and clubbing. Review and possible mechanisms. *Texas State Journal of Medicine*, 1965, **61**, 620.
7. Young, J. R. Ulcerative colitis and finger clubbing. *British Medical Journal*, 1966, **1**, 278.

Chapter 12

Nail Changes in General Medical Disorders and other Systemic Disorders

Increasing attention is now being paid to the careful scrutiny of the fingernails as part of the physical 'work up'. This is especially important in patients presenting for the first time to family physicians or internists; only a decade ago the routine and thorough examination of the fingernails did not feature in many of the books and manuals on methods of clinical examination. Until recently only dermatologists regularly examined the fingernails. Such is not the case in the mid-1980s.

The ready accessibility of the nails, the availability of ten fingernails to improve the statistics of bias and erroneous observations and their use for comparison and the increasing cost of laboratory tests, have all emphasised the value of the physical examination. So has the use of the two-handed medical handshake at the time of greeting. This is increasingly and widely taught to all young medical assistants and allows a smooth and almost undisturbing observation of the nails.

This routine examination of the nails should always be carried out within the framework and as part of *The Normal Physical Examination*. As a method for eliciting information and as a means of teaching deductions from observations, it is at least as useful as the pulse taking. Because it reflects past and present health, it allows a wider and more varied series of deductions; including skin and integument, state of nutrition and health, attitudes, personality and occupation.

In over a thousand consecutive admissions over the age of 60 years to the general medical ward of a teaching hospital, seldom did examination of the fingernails in a teaching ward-round setting fail to reveal valuable information about that person's make-up and background. Furthermore, 28 per cent had abnormal changes in the nails useful for both teaching of trainee doctors and as a source of additional data about the systemic disorders present.

Many older patients have abnormalities of the nails such as those described in other chapters – Beau's lines, koilonychia, clubbing, loss of lunula due to low serum albumen and other interesting coloured nails. It is also known that collagen disorders such as scleroderma and dermatomyositis may give significant nail changes.

The following group of nails is to emphasise the wide variety of general medical disorders from splinter haemorrhages (Miller & Vaziri, 1979) to thyrotoxicosis and Addison's disease which may give characteristic changes in the nails.

It is necessary that if during the normal examination an abnormality of the nail is seen, a cause or explanation for that change should be sought. So often elderly patients who have slowly developing nutritional deficiencies may accept as a normal evolution of old age changes which are to a trained observer clearly abnormal. The real tragedy is if their primary care physician does the same. This may occur either because of a failure to make the correct deductions from the right observations, or a failure to examine the nails. Nothing is more encouraging to doctor and patient than a wide range of correct deductions being made in the first minutes of clinical examination or greeting.

Cardiovascular and circulatory

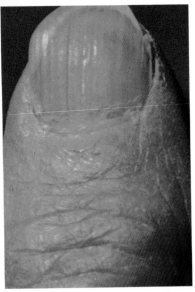

228

229

228 Valuable nail signs. Index finger nail from a 76-year-old widow with chronic and repeated episodes of left ventricular failure. She had become depressed, as manifested by the poor manicure at base of nail and the nails are slightly cyanosed. Warm skin temperature on hand-shake suggests this is central cyanosis and suggests some pulmonary oedema at the lung bases. The flattened front half of the nails may also suggest poor diet which was indeed the case because of digoxin toxity. Note also lack of lunula and some 'beading' in the normal longitudinal ridges of the nails.

229 Repeated illness. This thumbnail comes from a 67-year-old man with repeated severe Stokes-Adams attacks.

230 Useful signs. Thumbnail from 73-year-old non-insulin-dependent diabetic male. Eight months ago suffered myocardial infarction with poor output and hypotension. Note onycholysis leading edge and 1 to 2 months of dystrophic growth. Then a further infarction (4-month Beau's line). Also long-standing diabetic proteinuria and lowered serum albumen.

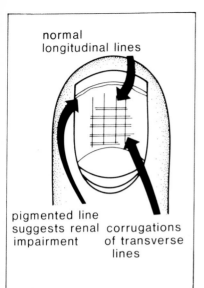

normal
longitudinal lines

pigmented line
suggests renal   corrugations
impairment   of transverse
                  lines

229 a

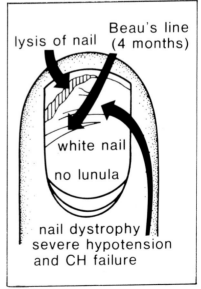

ATHEROMA NAILS

● Slow growth
● Opacity
● Brittleness

231

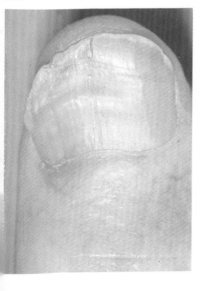

230

Beau's line
(4 months)

lysis of nail

white nail

no lunula

nail dystrophy
severe hypotension
and CH failure

230a

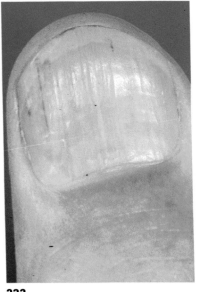
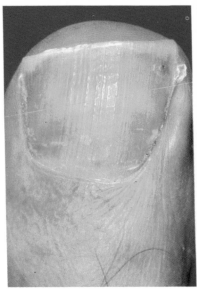

232   **233**

232 Widespread atheroma. The thumbnail of a 79-year-old male with onset of left middle cerebral artery thrombosis. Beau's lines and longitudinal ridges seen, left and right sides show remarkably similar changes.

233 Beading toenail. Great toenail from 73-year-old non-insulin-dependent diabetic with repeated heart troubles.

234 Nail shedding secondary to atheroma. An elderly male with gross aortic arch atheroma and ischaemia in hand.

235 and 236 Pigmented beading. Thumbnail (right) and toenails from a 35-year-old woman. These are from the right side of the body of a patient who had a hemispherectomy 25 years ago. Note pigmented longitudinal ridges in thumb and spasticity of toes.

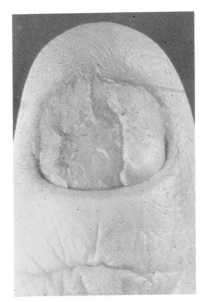

234

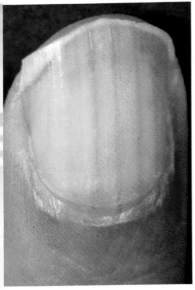

235

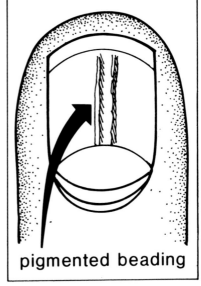

pigmented beading

235a

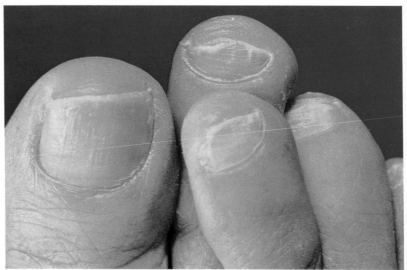

236

237 Chronic left heart failure. 77-year-old retired accountant with chronic left ventricular failure. Note excellent lunula, bluish nails and early onycholysis. 'Heaped up' posterior nail fold.

238 Raynaud's disease. Arrow shaped nails due to lateral onycholysis in elderly woman with many years of Raynaud's disease.

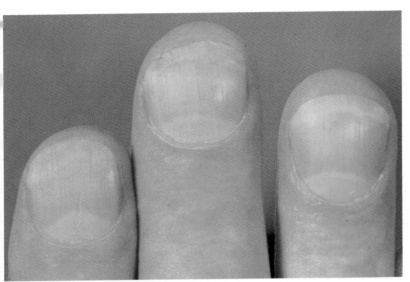

237

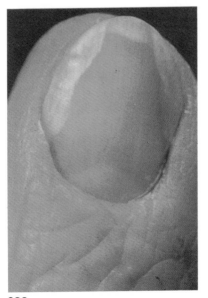

238

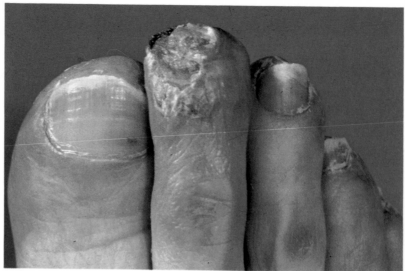

239

239 Severe Raynaud's. Very severe Raynaud's phenomena in feet with failure to respond to treatment. Secondary to autoimmune disease – note changes in skin colour.

240–243 Photographs of nails of patient with renal failure and hyperparathyroidism showing terminal bone absorption in the last digit of each finger.

Kidney

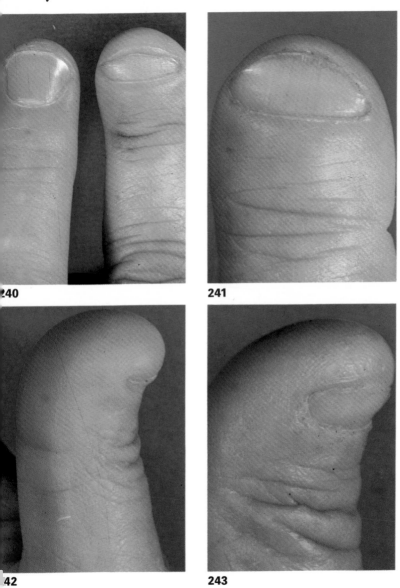

244

244 Early kidney failure. Interesting thumbnails in a 38-year-old Caucasian woman with diabetes for 27 years and now slowly developing renal failure.

245 Diabetic kidney failure. Fingers of 47-year-old man with insulin dependent diabetes mellitus for 22 years. Early visual failure due to retinopathy (note poor manicure) and protein losing nephropathy, also blood urea raised.

246 Multiple changes in kidney failure. Middle-aged woman with progressive renal failure and rising blood urea. Note poor appetite has led to some malnutritional changes with koilonychia and loss of lunula. Pigment below yellow line also seen in kidney failure and some early onycholysis of nutritional origin.

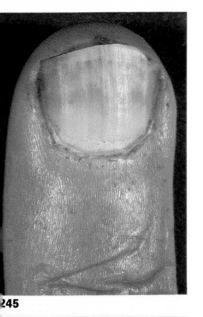

245

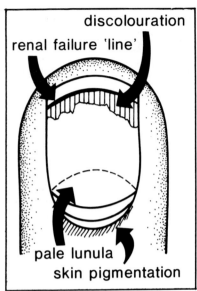

245a

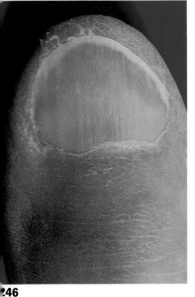

246

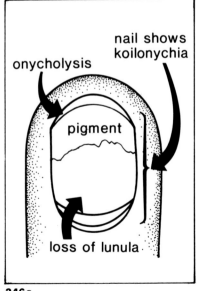

246a

151

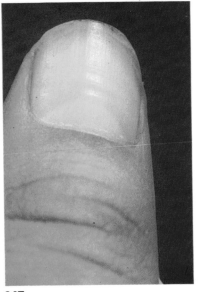

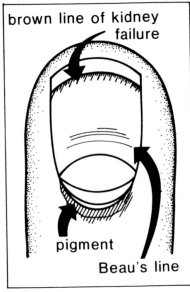

247 **247 a**

247 Maori with kidney failure. Thumbnail from 15-year-old boy who had a renal transplant 1 month previously. Note marked growth arrest line and pigmentation at base of nail.

248 Liver damage. Fingers from woman with alcoholic liver disease. Note infection and increased blood flow at base of nail. Also loss of lunula and onycholysis.

249 Terry's half and half nail. From same middle-aged woman with alcoholic liver disease and low serum albumen.

250 Muehrcke's white bands. Liver disease, early cirrhosis in a 38-year-old alcoholic woman with loss of lunula and white bands. Serum albumen at lower limit of normal.

Gastrointestinal and liver

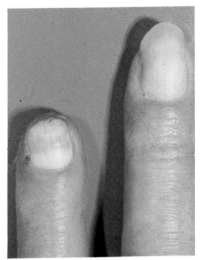

248

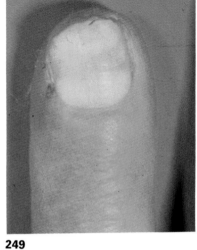

249

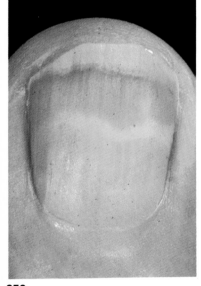

250

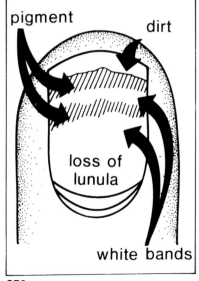

250a

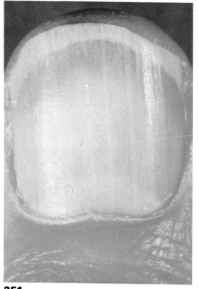 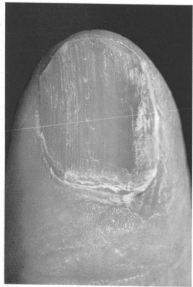

251 **252**

251 Pancreatic damage. Diabetes in a middle-aged lawyer. Pale nails and poorly defined lunula but serum albumen normal.

252 Nutritional. General nail dystrophy in middle-aged woman with upper gastrointestinal disease. Note lateral onycholysis, opacity and granularity of nails. Lunula is reduced. Poor manicure due to chronic ill health.

253 Acromegalic fingers. Untreated acromegaly in a 32-year-old single school teacher with chronic fungus infection in middle and index fingernails. Note statistical association between acromegaly with carbohydrate intolerance and high incidence of fungal infection of nails.

254 Acquired brachyonychia. Characteristic shrinking of terminal phalanx in secondary hyperparathyroidism.

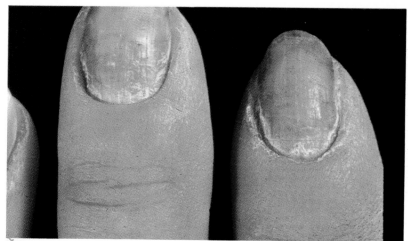

253

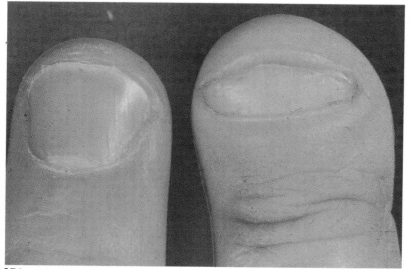

254

255

256

255 Elderly thyrotoxic. Thumbnail from hospitalised 80-year-old woman with congestive heart failure on basis of thyrotoxicosis. Note change in skin, good lunula, pigment in skin at nail base and 'beading' present in the longitudinal ridges usually seen in patients of this age.

256 Thyrotoxicosis. Marked 'beading' or 'raindropping' in an elderly thyrotoxic male. Picture at diagnosis. Also called 'sausaging'.

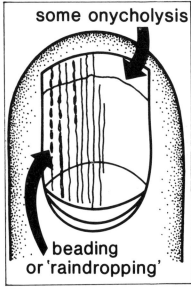

some onycholysis

beading
or 'raindropping'

256a

Scleroderma

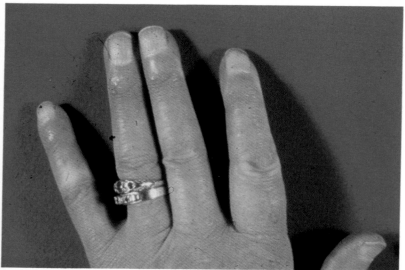

257

257 Scleroderma hands. Nails taper to a point with tethering of tissue in scleroderma.

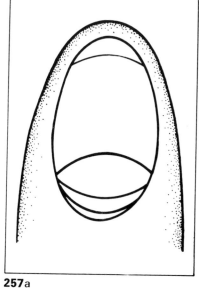

257a

258

259

258 Scleroderma. 41-year-old man with scleroderma presenting as early renal failure. Note extreme ridging of nails, poor manicure and loss of lunula. Tips of digits are narrowed.

259 Extrusion under nail. Nails from an elderly woman with severe scleroderma.

260 Scleroderma. Extréme tapering of fingers but nails still intact in more severe scleroderma.

261 Scleroderma and renal failure. Non-specific ridges but also loss of lunula and poor manicure.

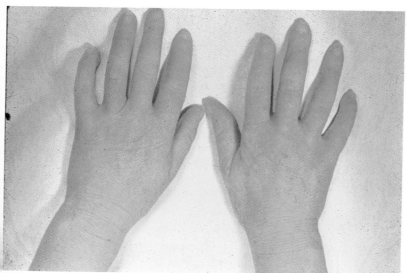

260

261

Blood and general

262 Bleeding under nail. Butazolidin-induced thrombotic thrombocytopenic purpura in a 71-year-old farmer being treated for arthritis of the hip.

263 Subungual haemorrhages. Elderly male with lymphoblastic stage of acute on chronic leukaemia. Subungual haemorrhages secondary to iatrogenic induced low platelets. Note also ground-glass appearance of the rest of the nail.

262

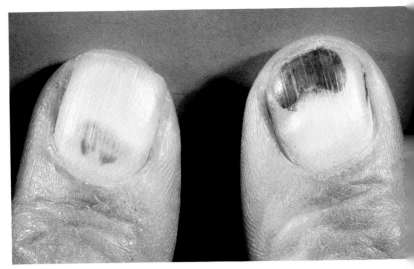

263

264 Close-up of subungual distal haemorrhages. Note bleeding in distal nail bed suggests spontaneous occurrence.

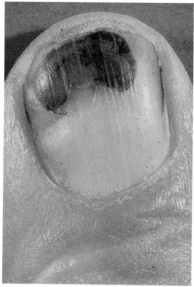

264

References

1. Gross, N. J. & Tall, R. Clinical significance of splinter haemorrhages. *British Medical Journal*, 1963, **2**, 1496.
2. Hamilton, E. D. B. Nail studies in rheumatoid arthritis. *Annals of Rheumatic Diseases*, 1960, **19**, 167.
3. Leyden, J. J. & Wood, M. F. The half and half nail of uremic onychodystrophy. *Archives of Dermatology*, 1972, **105**, 591.
4. Miller, A. & Vaziri, N. D. Recurrent atraumatic subungual splinter haemorrhages in healthy individuals. *South African Medical Journal*, 1979, **72**(11), 1418.
5. Norton, L. A. Nail disorders. A review. *Journal of the American Academy of Dermatologists*, 1980, **2**(6), 451.
6. Quenneville, J. G. & Gossard, D. Subungual-splinter haemorrhage an early sign of thromboangiitis obliterans. *American Journal of Diseases of Childhood*, 1981, **135**(4), 383.
7. Ridley, C. M. Pigmentation of fingertips and nails in vitamin B12 deficiency. *British Journal of Dermatology*, 1977, **97**(1), 105.
8. Urowitz, M. B., Gladman, D. D., Chalmers, A. & Ogryzlo, M. A. Nail lesions in systemic lupus erythematosus. *Journal of Rheumatology*, 1978, **5**(4), 441.
9. Young, J. R. Ulcerative colitis and finger clubbing. *British Medical Journal*, 1966, **1**, 278.

Chapter 13

Nail Changes Associated with Intellectual Handicap and Congenital Disorders

This is a less common group of disorders of the fingernails, and also of the toenails. The study of light microscopic changes in the chromosomes over the last decade and the publication by J. M. Robert and colleagues (1977) have stimulated an interest in these disorders. The classification of Robert with the recognition of monosomy or trisomy of a number of different chromosomes in association with characteristic nail findings has stimulated an interest in nail appearances in all young people with intellectual handicap and learning difficulties. The Robert classification is also available in an adaptation by Baran (1981) which is perhaps more accessible to readers in English.

For the general practitioner or specialist in internal medicine, the actual recognition of:

Micronychia
Incurving or short fifth fingers
Short nails
Nail en racquette
Lack of lunula
Fused nails
Syndactyly

Nail dystrophy with thickening, excess ridging or shedding should always alert one to the need for chromosome studies. Pachonychia congenita with hyperkeratoses and the nail patella syndrome have long been known. More recently this field of study is rapidly expanding as abnormal nails prove to be markers for:

Chromosomal abnormalities, e.g., Down's syndrome, trisomy-21.
Hereditary deformities with normal karyotype, e.g., Apert's syndrome.
Genetic disorders of metabolism with one or more inherited defect, e.g., homocysteinuria.

Down's syndrome *(Trisomy 21)*

265 Trisomy 21. Down's syndrome, trisomy 21. Thumbnail within normal limits whereas often short in Down's syndrome.

266 Normal length nails. Fingernails from Down's syndrome show abnormal longitudinal ridging and lack of manicure seen in association with mental deficiency.

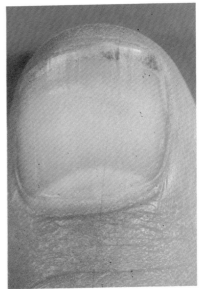

265

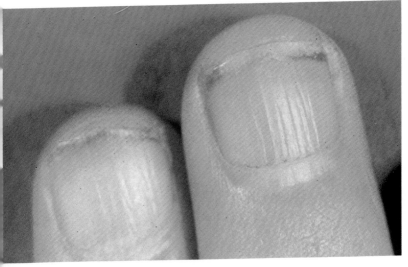

266

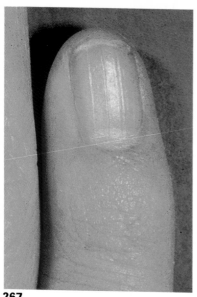

268

267

267 'Old' nail for 18 years. Incurving short 5th finger as seen in trisomy 21. Note that the nail is also abnormally narrow with longitudinal ridges.

268 Common brachynychia. Usual micronychia in trisomy 21 with normal adult of the same age for comparison.

269 Brachynychia thumb. Trisomy 21 with profound mental deficiency. Note micronychia of thumbnail. No epicanthic fold changes but marked laxity of ligaments.

270 **and** 271 Nail dystrophy. Trisomy 21 with unilateral arm, oedema and marked nail dystrophy. Other side shows usual appearances in Down's syndrome with short thumbnail and incurving 5th finger. On unaffected side poor development of the lunula can be seen.

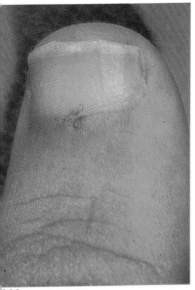

269

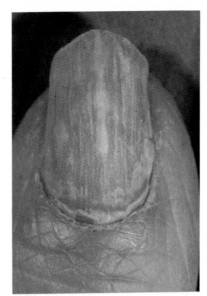

270

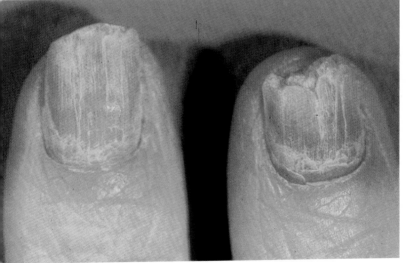

271

165

272

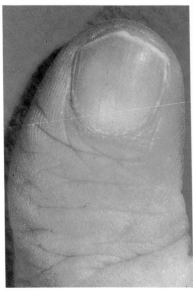

273

272 Down's syndrome. Short nail, abnormally convex and no lunula.

273 Trisomy 21. Incurving 5th finger. Poor nail cutting and absent lunula.

274 Trisomy 21 – 5th finger. Nails typical of Down's syndrome. Note incurving 5th finger with short broad nail which also shows micronychia and loss of lunula.

275 and 276 Elderly Down's. Nails from a long-surviving 64-year-old man with trisomy 21. 275 Thumbnail shows short broad and convex nail with poorly developed lunula. 276 First finger also short broad nail – some thinning of nail near free edge.

277 Probable trisomy 21. Microcephaly and severe mental retardation. Nail dysplasia and lysis. No chromosome studies yet available.

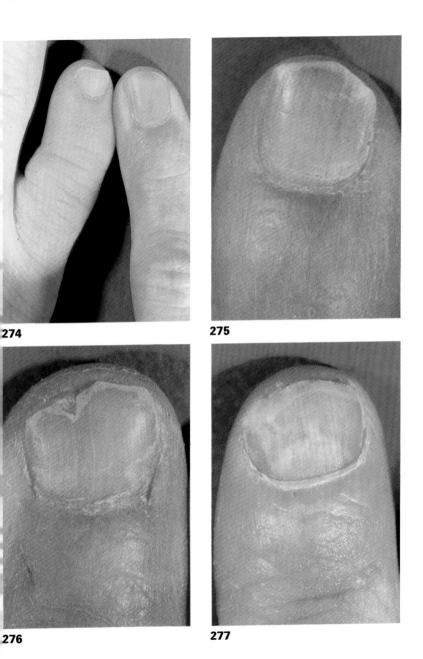

274

275

276

277

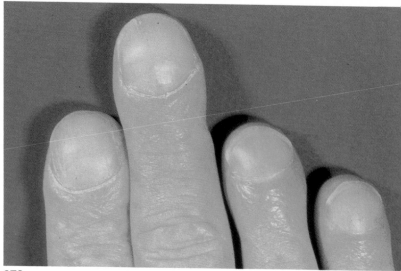

278

278–280 Intellectual impairment, trisomy 21, with Eisenmenger's syndrome and pulmonary stenosis. Note elongated middle finger and marked clubbing which has dominated nail development. **278** View of elongated finger. **279** Normal adult 1st finger from a person of same age showing enlargement of fingertip. **280** Drumstick clubbing and cyanosis marked. Lunula unduly large and unduly pale.

281 and 282 Trisomy 21 in adult patient of moderate intellectual impairment. Minimal degree of psoriasis also present with right 5th finger affected. Left 5th finger shows normal Down's syndrome features: incurving of finger with reduced lunula, convexity and increased longitudinal ridges. As well as psoriasis, middle finger also shows old injury.

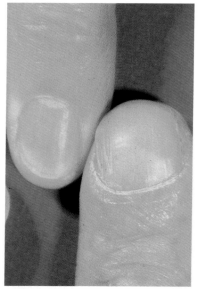

279

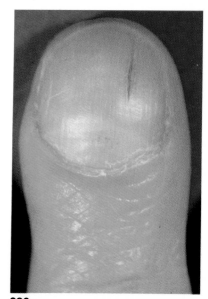

280

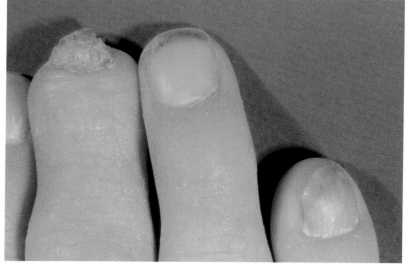

281

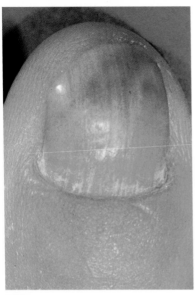

282

282 Close-up psoriatic nail.

283 Monosomia 4. Totally absent lunula in young woman with severe intellectual impairment, tapered fingers, lax ligaments and monosomia 4. Note appearance of 'splinter' haemorrhages which are actually petechial dots.

284 ?Trisomy 7. Possibly trisomy 7 with severe mental retardation with dysmorphia of skull. Some hypotoma but marked incurving of 5th finger and narrow convex nail with no lunula.

285 Profound retardation – trisomy 8. Minor changes in hands, incurving 5th finger and marked dysplasia of nails.

286 Monosomia 9. Monosomia 9 with severe psychosis and severe mental retardation. Nails widened with some lysis.

Chromosome and other syndromes

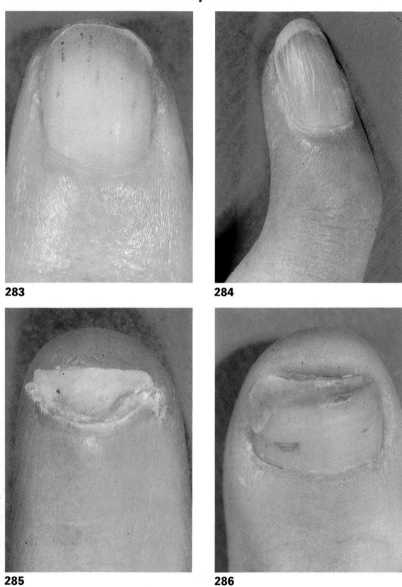

283

284

285

286

287

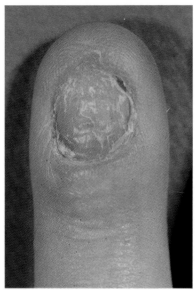
288

287 Quadrosomy X. Nails from middle-aged male with moderate retardation and double Klinefelter's syndrome. Patient shows quadrosomy X with mosaicism and incurving 5th finger. Thumbnails short, rounded and convex.

288 Gross nail dystrophy. Dyskeratosis congenita. Repeated shedding of nails since birth in a young intellectually impaired woman.

289 Cri Du Chat syndrome. Note nails are short, slightly convex with longitudinal ridges unusual in a 20-year-old. Shown with normal nails from female of same age.

290 Wide 'flared' nails. Microcephaly, profound mental retardation. Short wide nails with no lunula and onycholysis. Chromosome status unknown.

291 Epiloia (tuberose sclerosis). Subungual fibromata seen in both middle and index fingers of the right hand in a woman mentally damaged from repeated fits.

292 Poor lunula. Microcephaly, mental retardation and short nails with marked longitudinal ridging and poorly developed lunula. ?Monosomia 4.

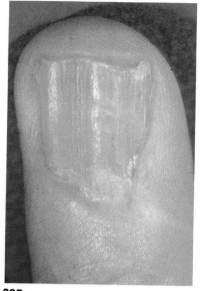

295

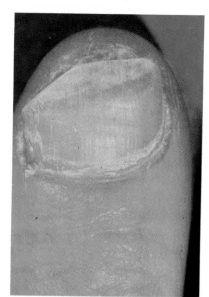

297

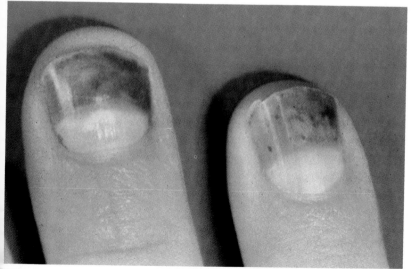

296

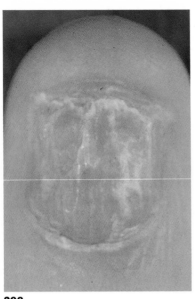

298 Dyskeratosis congenita. Also mental deficiency and microcephaly.

298

299–302 Noack's syndrome. Noack's syndrome with polysyndactyly and grossly abnormal nails. ?Trisomy 13.

299 Convex micronychia in fused digits.

300 Extreme micronychia of thumb.

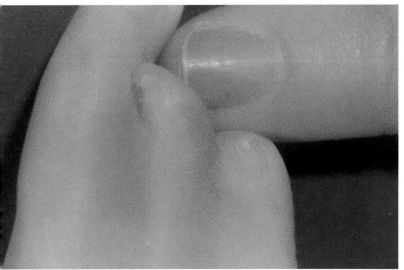

299

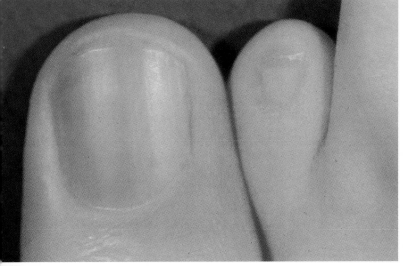

300

177

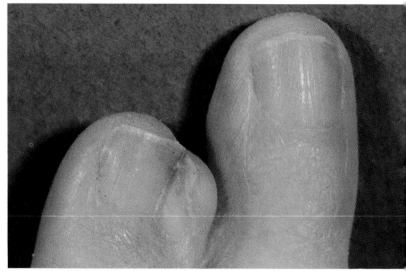

301

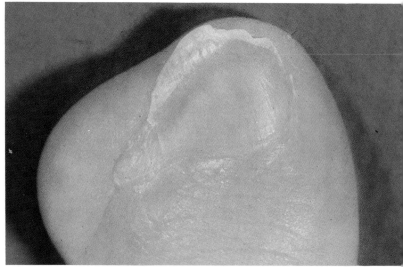

302

178

301 The 'normal' nails show loss of lunula and convexity.

302 Gross abnormality of thumb of right hand.

303–305 Apert's syndrome. Oxycephaly, mental impairment and hypertelorism with syndactyly. Note the fusion of digits results in a single large nail (normal adult hand alongside). Thumb takes on a 'flipper' like role, convex and without lunula.

303 Broad fused single nail from 4 fingers.

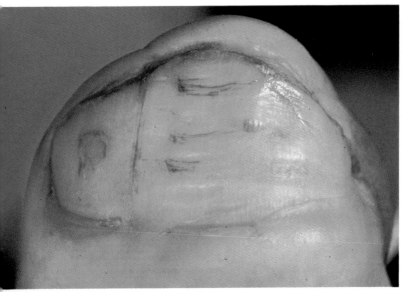

303

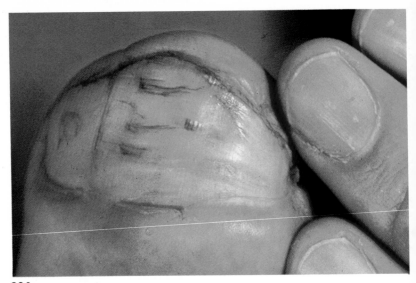

304

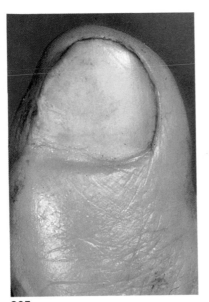

305

304 Syndactyly of left hand (normal adult fingers alongside).

305 Close-up of thumb to show convexity and absence of lunula. Contrary to reports there is no real atrophy of nail, but a fusion of all.

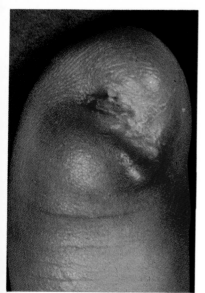

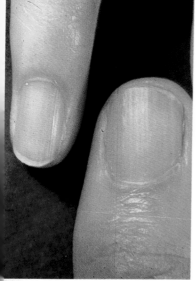

306

307

306 Amniotic nail syndrome. Absent nail since birth in 5th finger due to reduced amniotic fluid. All other nails normal.

307 **and** 308 Marked arachnodactyly. Intellectual impairment – arachnodactyly and homocysteinuria in an adult woman with long tapered fingers and nails. (A) Note incurving of 5th finger and length of digits. (B) Micronychia and elongation of nails shown well in the 5th finger compared to normal.

308

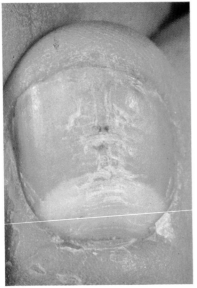
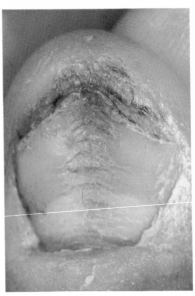

309 **310**

309 and 310 Severe intellectual impairment from birth with central nail dystrophy. 309 Left thumbnail. 310 Right thumbnail resembles Leclercq canaliformis changes but is almost certainly due to repetitive trauma.

References

1. Baran, R. in *The Nail*, Pierre, M. (ed), Churchill-Livingstone, London, 1981, p.65.
2. Bassett, M. R. H. Trois formes genotypiques d'ongles courts: le pouce en raquette, les doigts en raquette, les ongles courts simples. *Bull. Soc. fr. Derm. Syph.*, 1962, **69**, 15.
3. Boxley, J. D. Pachyonychia congenita and multiple epidermal hamatomata. *British Journal of Dermatology*, 1971, **85**, 298.
4. Cole, H. N., Raushkolb, J. E. & Toomey, J. Dyskeratosis congenita with pigmentation, dystrophia unguis and leukokeratosis oris. *Archives of Dermatology*, 1930, **21**, 71.
5. Franzot, J., Kansky, A. & Kav'Ci'C, S. Pachyonychia congenita (Jadassohn-Lewandowsky syndrome). A review of 14 cases in Slovenia. *Dermatologica*, 1981, **162**(6), 462.

6. Goodman, R. M. & Cuppage, F. E. The nail patella syndrome. Clinical findings and ultrastructural observations in the kidney. *Archives of Internal Medicine*, 1967, **120**, 68.

7. Hazelrigg, D. E., Duncan, C. & Jarratt, M. Twenty nail dystrophy of childhood. *Archives of Dermatology*, 1977, **113**, 75.

8. Kitayama, Y. & Tsukada, S. Congenital onychodysplasia. Report of 11 cases. *Archives of Dermatology*, 1983, **119**(1), 8.

9. Nevin, N. C., Thomas, P. S., Calvert, J. & Reid, M. M. Deafness, onycho-osteodystrophy, mental retardation (DOOR) syndrome. *American Journal of Medical Genetics*, 1982, 13(3), 325.

10. Pfeiffer, R. A. The oto-onycho-peroneal syndrome. A probably new genetic entity. *European Journal of Pediatrics*, 1982, **138**, 317.

11. Robert, J. M., Planchu, H., Giraud, F. & Matei, J. E. In *Genetique et Cytogenetique Cliniques, Paris*, Flammorian Medicine Sciences, 1977.

12. Ronchese, F. The racket thumb nail. *Dermatologica*, 1973, **146**, 199.

13. Sparrow, G. P., Samman, P. D. & Wells, R. S. Hyperpigmentation and hypohidrosis. *Clinical and Experimental Dermatology*, 1976, **1**, 127.

14. Stieglitz, J. B. & Centerwall, W. R. Paronychia congenita (Jadassohn-Lewandowsky syndrome): a seventeen-member, four-generation pedigree with unusual respiratory and dental involvement. *American Journal of Medical Genetics*, 1983, **14**(1), 21.

Chapter 14

Coloured Nails or Chromonychias

The colour of the nail is dependent upon the following:

Thickness of the nail plate and its transparency.

The nature of the blood and its composition, e.g., a lack of red cells or anaemia will give pale nails and a large amount of reduced haemoglobin will give blue nails.

The state of the blood vessels: If in spasm, the arterioles will deliver less blood to the nail bed. The special control of capillary shunts by the circulating kinins in association with autonomic control is especially important to the nail bed as its arterial supply in part must first traverse these capillary shunts and digital pulp plexus before returning to the nail bed afterwards. This is particularly important in the toenails which thus show early signs of peripheral vascular disease with opacity, thickening due to slow forward growth and dystropias.

The main abnormalities in colour require an excellent light source, preferably sunlight. Changes in colour can be divided into three main groupings: Those dependent upon whether the chromogenic agent is being applied from outside, as in certain occupations; secondly, those other dyschromias arising from within the nail bed or nail plate and these are usually classified as endogenous. Of these, white nails or leukonychia is the common and most important. The more general types are set out in the associated table.

In all coloured nail syndromes, a meticulous history of occupation, drug and medication ingestion and personal and cosmetic habits is vitally necessary.

1.0 **White nails or leukonychias.**
      (a) *Nail plate.*
1.1 Total leukonychia.
1.2 Partial leukonychia.
1.3 Striated leukonychia.
1.4 Congental leukonychia.
      (b) *Subungual leukonychia.*
1.5 Terry's white nails.
1.6 Muehrke's white bands.
1.7 Uraemic 'half-and-half' nails.
1.8 Dermatology disorders, e.g., psoriatic white nails.

2.0 **Exogenous chromonychias.**
2.1 External applications, e.g., silver nitrate, gential violet, etc.
2.2 Cosmetics.
2.3 Occupational, i.e., hairdressers and those working with french polish and varnish.
2.4 Trauma.

3.0 **Endogenous dyschromias.**
      (a) *Toxic and therapeutic.*
3.1 Poisons such as lead, silver, arsenic, etc.
3.2 Drugs such as phenothiazines, anti-malarials, tetracyclines.
3.3 Anti-mitotic drugs.
3.4 Drugs causing platelet changes and haemorrhages.
      (b) *Systemic disorders, infections
         and nail bed pathologies.*
3.5 Blood stream colours such as bilirubin and carotenaemia, etc.
3.6 The yellow nail syndrome associated with lymphatic diseases.
3.7 Pigmented nails associated with endocrine disorders.
3.8 Other cardiovascular, metabolic and congenital diseases such as Wilson's disease, polycythaemia, beri beri, anaemia, other hypoalbuminaemias, etc.
3.9 Infections. Examples would be the black and brown nails seen in onychomycoses, the green nails of pseudomonas, yellow nails of scopulariopsis, white nails of aspergillosus, etc.

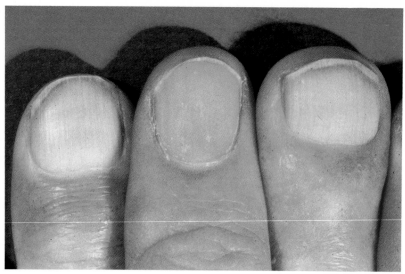

311

311 Leukonychia – tinge of cyanosis. Retarded 60-year-old with mitral incompetence and congestive heart failure. Drug treatment has led to a mild dermatitis. Note bluish fingernails when compared with normal adult of same age. Nail with relative loss of lunula and bluish appearance of nail plate.

312 Leukonychia. Opaque white nails seen in a male in his 30s with alopecia totalis.

313–315 Chronic liver disease. Chronic persistent hepatitis (non-alcoholic) in post-menopausal woman. Note whiteness of all nails (serum albumen only 2.8 g/l and marked palmar erythema.

312

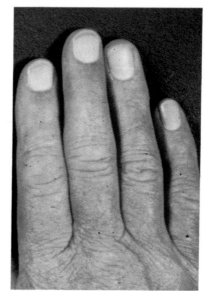

313

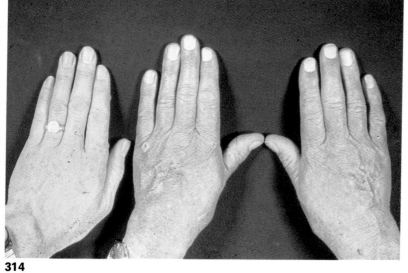

314

187

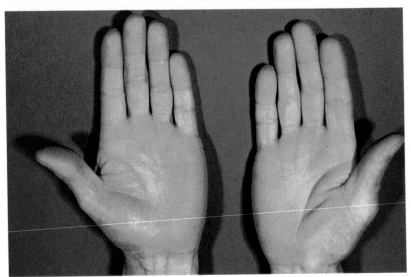

315

316 Variable leukonychia. Hand from patient with marked renal failure and low serum albumen showing 'half and half' nails.

317 Leukonychia striata in an elderly woman, probably due to repeated trauma to nail fold.

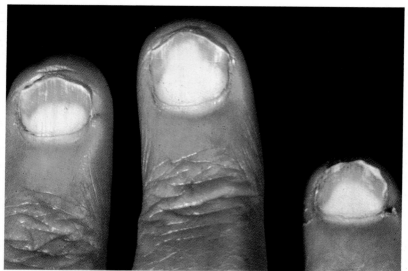

316

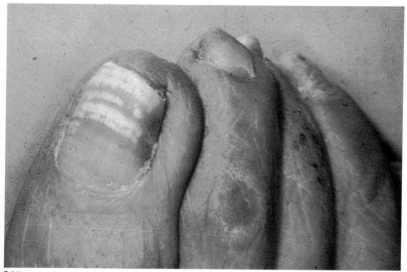

317

189

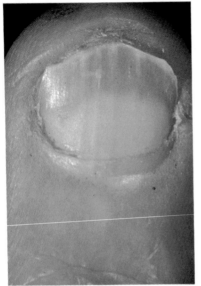

318

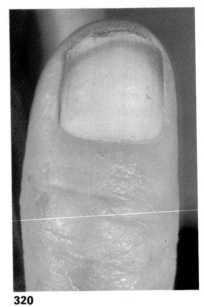

320

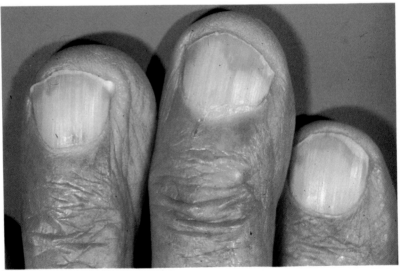

319

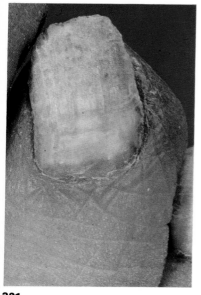
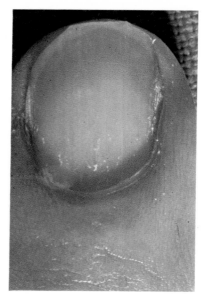

318 'Half and half' nail. A good example of Terry's 'half and half' nails. This time due to renal failure.

319 Marked leukonychia. Severe total leukonychia in elderly patient with renal failure. Note small area of onycholysis and pigmented skin of organ failure.

320 Multiple changes with white nails. Coloured nails in middle-aged male with chronic heart failure secondary to valvular disease. Cyanosis is present, slight jaundice and the brown line under the free edge is secondary to nitrogen retention. The renal failure appears related to the low output state in this neglected patient. Note manicure.

321 Chromatodystrophy. Yellow-white chalky nails in a patient with Down's syndrome.

322 Unusual distal colouring. Terminal or distal nail discolouration in renal failure. The more usual nail signs are a thin dark brown line behind the yellow line or loss of lunula.

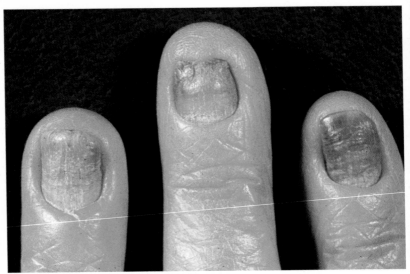

323

323 Yellow nails. Yellow nail syndrome seen in lymphatic abnormalities, especially thoracic and intestinal lymphangiectasia (Samman & White, 1964; Hiller *et al.*, 1972).

324 Yellow nail. Congenital abnormality of lymphatics with yellow nail dystrophy and secondary infection.

325 Yellow nails in feet. Yellow nail syndrome due to lymphatic hypoplasia in a 64-year-old farmer presenting with gross leg oedema and supposed diagnosis of congestive heart disease.

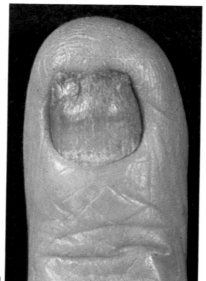

324

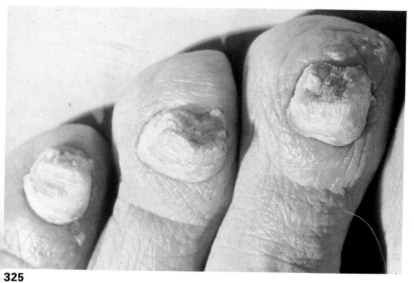

325

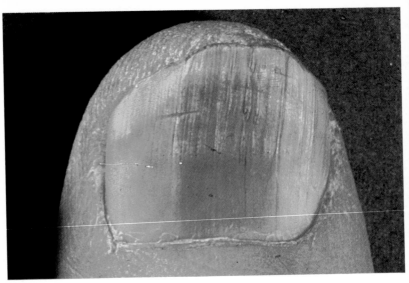

326

326 Yellow-brown nails. Brown nails secondary to thyrotoxicosis in a middle-aged male of Caucasian descent. Note also marked distal onycholysis, growth arrest areas and prominent longitudinal nail ridges, also absent lunula.

327 Green nail syndrome. Pale green nail not in this case due to *Pseudomonas*. Likely exogenous pigment in damaged nail.

328 Green nail. Chronic *Pseudomonas* infection in toenail. Characteristic chronicity.

329 Green-yellow nail. Green toenail seen in a man in his 30s who sustained a cricket ball injury to the end of the great toe as a teenager. No lymphatic abnormalities and no fungus in nail scrapings.

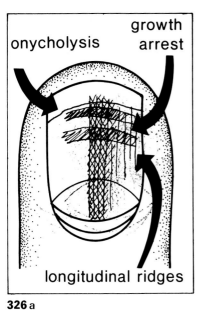

326 a

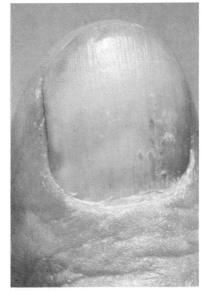

327

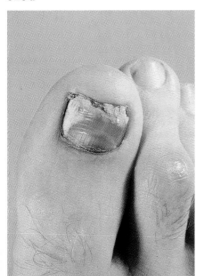

328

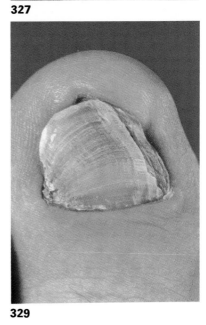

329

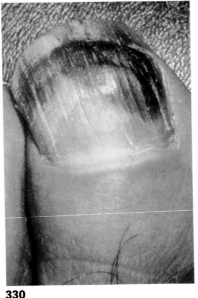

330

331

330 Acute *Pseudomonas chromonychia.*

331 Thinning of nail – pink nail. 60-year-old man with heart failure and very thin nails: thin nail plate presenting as pink nail. Note lunula just present.

332 Further 'pink nail' due to thinning. Severe distal nail dystrophia in a 37-year-old man with 'controlled' ulcerative colitis. Amyloid disease present on biopsy, etc. ?Amyloid nails as nail biopsy +.

333 Pink nail syndrome. 64-year-old widow with marked nutritional disorder. Failure to eat adequate B and C vitamins as well as iron and protein gives 'pink' nail syndrome. Nails flattened, thin, with loss of lunula and measured serum levels of iron, thiamine and vitamin C below lowest normal values.

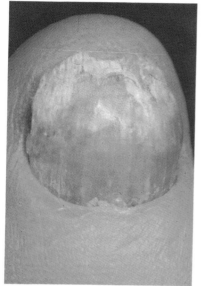

332

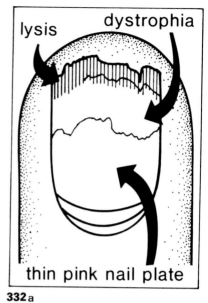

lysis   dystrophia

thin pink nail plate

332a

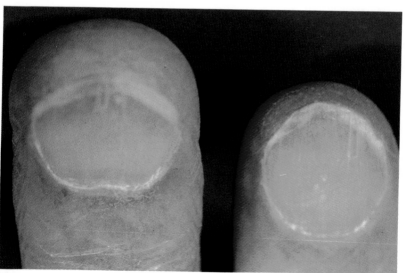

333

197

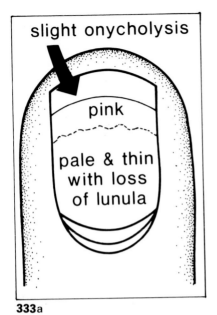

333a

334 Pink with striae. Loss of lunula in depressed 67-year-old widow. History of eating no protein but vegetables only. Serum albumen 3 g/l. One small Beau's line and normal longitudinal striations.

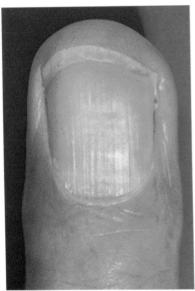

334

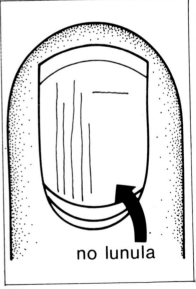

334a

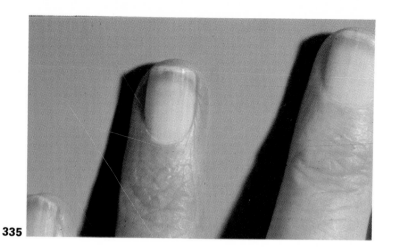

335 Mauve nails. Mauve nails, combination cyanosis, whiteness and lack of lunular in a renal failure patient give this appearance. Note renal failure line under the free edge.

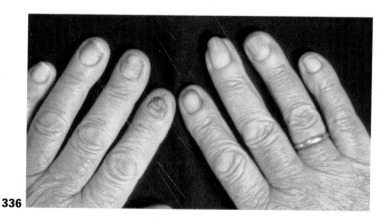

336 Dark mauve. Congestive heart failure which is chronic with cyanosis in a woman with Raynaud's disease of the hands. Note fungus of the nails.

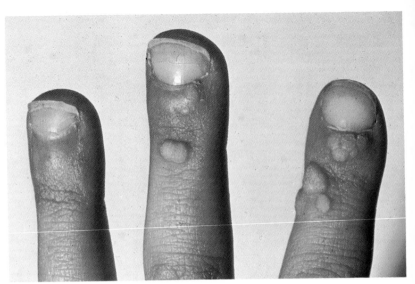

337

337 Blue nails in child. Blue nail syndrome in 10-year-old with colour change
and hypertrophied terminal pulps since birth. Other congenital bone deformities
but no lung or heart disease. Note also numerous viral warts.

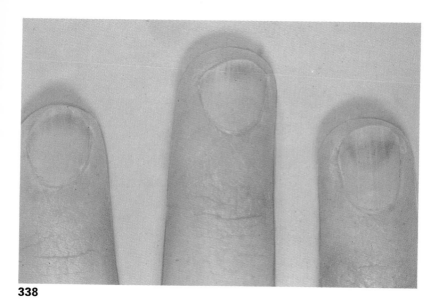

338

338 Pale blue with pigment. Pigmentation on distal or leading edge of several nails in a young woman. Probably due to drugs. ?Phenothiazines.

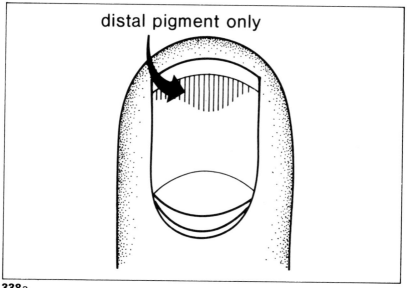

distal pigment only

338a

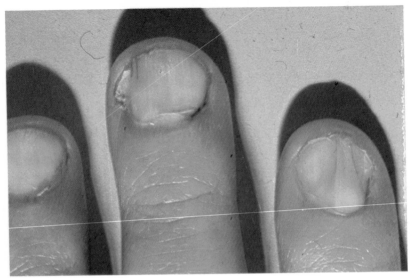

339

339 Pigment lines. Nail damage due to fish packing with poor manicure of nails in a factory worker. Note also pigment lines in two nails.

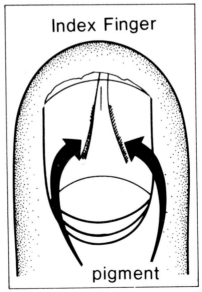

339a

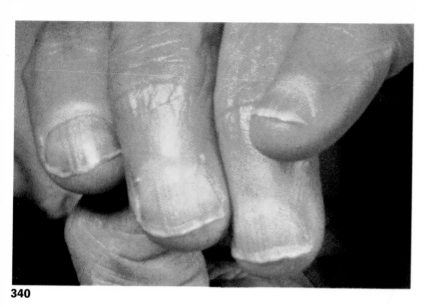

340 'Lilac' nails. 73-year-old woman with congestive heart failure secondary to coronary heart disease with poor appetite due to digoxin toxicity. Note 'lilac' rather than blue colour. Also note poor care of nails.

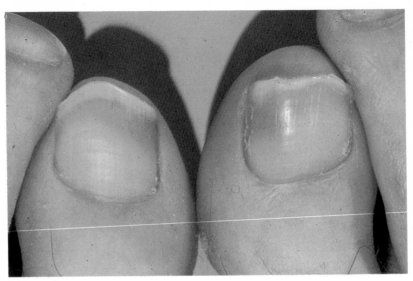

341

341 Kidney failure pigment. Brown lines of renal failure. Here seen in fingers but better seen in toes. This brown line spreads further proximally as a discolouration.

342 General pigmentation. General pigmentation in thumbnails of 70-year-old Chinese woman with 6 months of unexplained vomiting and weight loss. Chronic stress, no measurable nutritional abnormalities and no pathology found to account for condition. Note normal lunula and area of distal onycholysis.

343 Psoriatic pigmentation. Black nails due to psoriasis but not on treatment and no obvious skin lesions. Note also marked 'pitting' of nails – resembles pottery 'salt glaze'.

344 Multifactorial pigmentation. Koilonychia (gross), some onycholysis and brown nails in middle-aged diabetic Caucasian male with thyrotoxicosis and malnutrition.

342

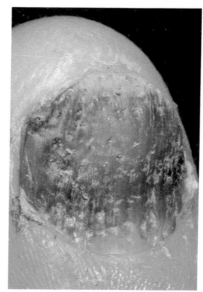

343

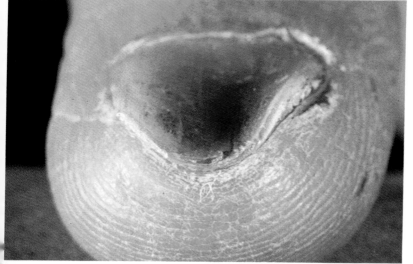

344

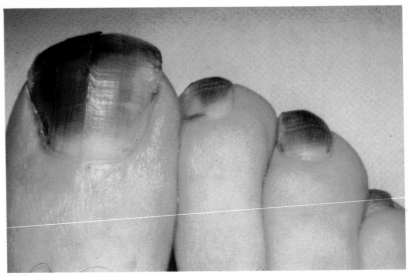

345

345 Pigment staining. Reddish-purple toenails due to staining with potassium permanganate (K_MNO_4).

346 Black arc pigment. Black proximal arc to nail plate, probably due to treatment of mild paronychia at nail base.

347 Brown line. Long narrow pigmented brown band in a male Caucasian. Addisonian with slow onset. A similar appearance is sometimes the result of a pigmented naevus at the base of the nail.

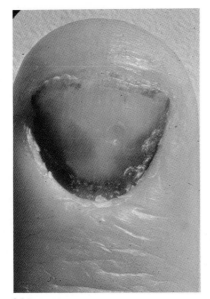

346

347

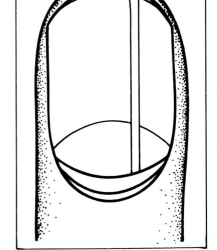

347a

207

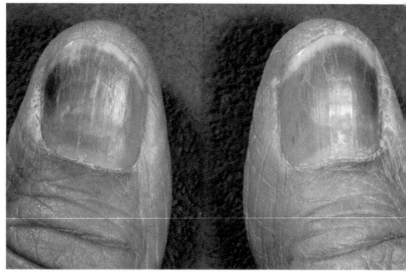

348

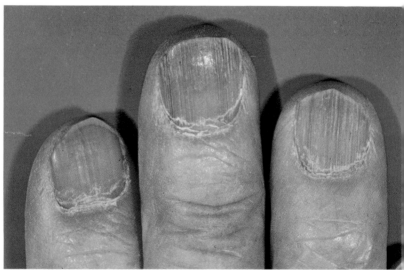

349

208

348 Pigmented nails – drugs. 62-year-old housewife with long-standing chronic obstructive airways disease and atalectasis of right lower lobe of lung. For two years on high dose therapy with continuous tetracyclines. Over previous 4 months had rovamycin and terramycin. No lung or adrenal failure and thyroid tests normal.

349 'Grey' nails. Black nails – in chronic heart disease. The patient is an elderly Chinese woman who noticed gradual onset of black nails and longitudinal ridges when she started on oral medication for her illness (diuretics and digoxin). No local medication.

References

1. Chapel, T. A. & Adcock, M. Pseudomonas chromonychia. *Cutis*, 1981, **27**, 601.
2. Hendricks, A. A. Yellow lunulae with fluorescence after tetracycline therapy. *Archives of Dermatology*, 1980, **116**(4), 438.
3. Ingegno, A. P. & Yatto, R. P. Hereditary white nails (leukonychia totalis), duodenal ulcer and gallstones. Genetic implications of a syndrome. *New York State Journal of Medicine*, 1982, **82**(13), 1797.
4. Jensen, O. White fingernails preceded by multiple transverse white bands. *Acta Dermato-Venereologica (Stockh)*, 1981, **61**(3), 261.
5. Jorizzo, J. L., Gonzalez, E. B. & Daniels, J. C. Red lunulae in a patient with rheumatoid arthritis. *Journal of the American Academy of Dermatology*, 1983, **8**(5), 711.
6. Kopt, A. W. & Waldo, E. Melanonychia striata. *Australian Journal of Dermatology*, 1980, **21**(2), 59.
7. Leyded, J. J. & Wood, M. G. The half and half nail of uremic onychodystrophy. *Archives of Dermatology*, 1972, **105**, 591.
8. Lindsay, P. G. The half and half nail. *Archives of Internal Medicine*, 1967, **119**, 583.
9. Loveman, A. B. & Fliegelman, M. T. Discolouration of nails. *Archives of Dermatology*, 1955, **72**, 153.
10. Luria, M. N. & Asper, S. P. Jr. Onycholysis in hyperthyroidism. *Annals of Internal Medicine*, 1958, **49**, 102.
11. Moore, M. & Marcus, M. D. Green nails: role of candida (Syringospore, Monilia) and Pseudomonas aeruginosa. *Arch. Derm. Suph. (Chic)*, 1951, **64**, 499.

12. Morey, D. A. J. & Burke, J. O. Distinctive nail changes in advanced hepatic cirrhosis. *Gastroenterology*, 1955, **29**, 258.
13. Satanove, A. Pigmentation due to phenothiazines. *Journal of the American Medical Association*, 1965, **191**, 263.
14. Segal, B. M. Photosensitivity nail discolouration and onycholysis. *Archives of Internal Medicine*, 1963, **112**, 165.
15. Shelley, W. B., Rawnsley, H. M. & Pillsbury, D. M. Postirradiation melanonychia. *Archives of Dermatology*, 1964, **90**, 174.
16. Stewart, W. K. & Raffle, E. J. Brown nail-bed arcs and chronic renal disease. *British Medical Journal*, 1972, **1**, 784.
17. Sutton, R. L. Transverse band pigmentation of fingernails after X-ray therapy. *Journal of the American Medical Association*, 1962, **150**, 210.
18. Suzuki, K., Uraoka, M., Funatsu, T., Sakaue, H., Onji, M., Ohta, Y. & Ishikawa, N. Cronkhite-Canada syndrome. A case report and analytical review of 23 other cases reported in Japan. *Gastroenterology Japan*, 1979, **14**(5), 441.
19. Valero, A. & Sherf, K. Pigmented nails in Peutz-Jeghers syndrome. *American Journal of Gastroenterology*, 1965, **43**, 56.
20. Young, W. J. Pigmented mycotic growth beneath the nail. *Archives of Dermatology*, 1934, **30**, 186.

Chapter 15

Tumours and Pterygia

Almost always tumours around the nail plate presented to the family physician are subsequently referred on to a dermatologist with an interest in the nails. Warts and small periungual fibromata around the nail folds are the most common simple tumours, followed by mucous cysts and exostoses. Pigmented naevi, glomus tumours, keratocanthomas and epidermoid inclusions are all rare and need specialised advice and surgical removal.

The very rare nail bed cancers such as epitheliomata and malignant melanomas may be seen perhaps once in a lifetime in elderly patients. A malignant melanoma may present as a small pigmented area secondary to a small localised paronychia. It may also appear as a junctional naevus.

If *any* pigmented lesion starts to spread up behind the nail or beyond the boundary of the nail plate, *urgent* referral is indicated for frozen section and possible removal of the nail and terminal digit.

Most pterygia arise from small penetrating injuries or disorders at the proximal end of the nail bed, deep enough to injure the matrix in a localised area. A prime example would be the small fibromata seen in epiloia where pterygia are commonly seen in association with the many small nail fold fibromata.

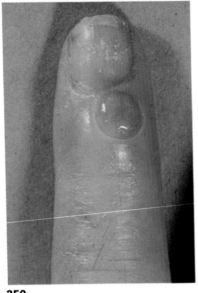

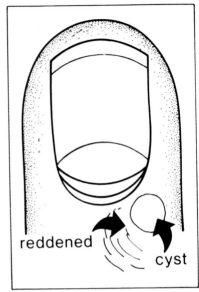

reddened

cyst

350

350a

350 Cyst of finger. A mucous or mucoid cyst of finger. Thought of as arising from the nail fold but actually arising from collagenous degeneration of extensor tendon.

351 Tumours and cysts. Mucous cyst over dorsum of the finger arising from degeneration of tendon.

352 Cyst with nail changes. Mucoid cyst of base of nail bed with associated distal changes in the nail.

353 Subungual fibroma. Nail excised away above a subungual fibroma.

354 Subungual exostosis of great toe.

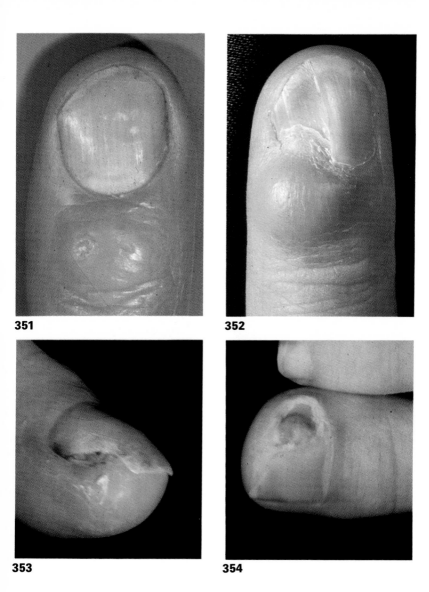

351

352

353

354

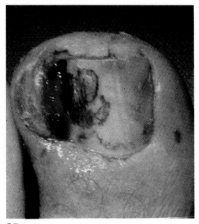

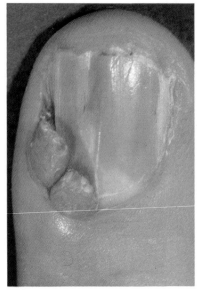

355

356

355 Nail bed tumour. Small mucoid tumour of great toenail bed (non-malignant).

356 Two subungual fibromata. Watch for other lesions as sometimes seen with epiloia and calcified cerebral vessels.

357 Pseudo tumours. Nails and hands in chronic tophaceous gout (extreme example).

358 Results of subungual fibroma. A more marked pterygium of two nails.

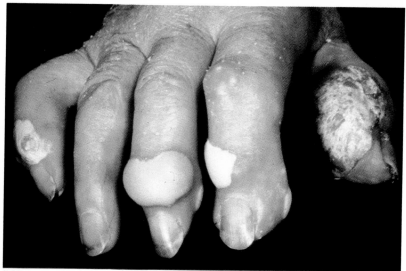

357

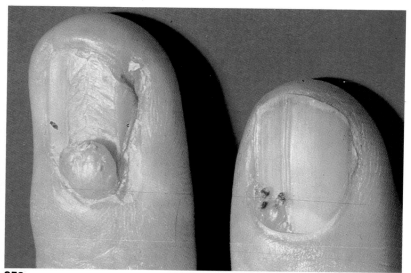

358

215

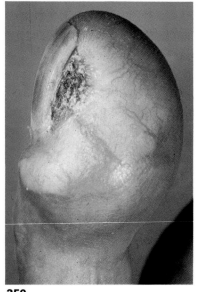

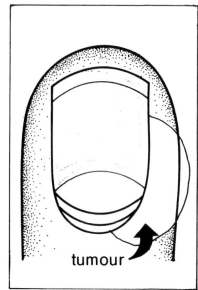

tumour

359   **359**a

359 Malignant tumour. Periungual malignant fibroma in a smoking pensioner. Note also 'beaking' of the nail.

360 Sarcoma of the nail bed.

361 Fibroma under central nail.

362 Small pterygium (= prow of ship). Early pterygium on one nail only. No obvious cause such as lichen planus was found.

360

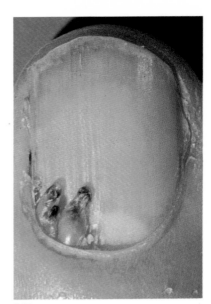

361

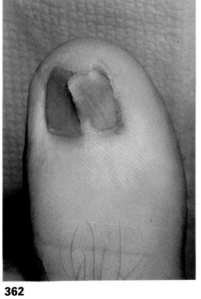

362

References

1. Arner, O., Lindholm, A. & Romanus, R. Mucous cysts of the fingers. *Acta Chir. Scand.* 1956, 3, 314.
2. Evison, G. Subungual exostosis. *British Journal of Radiology*, 1966, 39, 451.
3. Halpern, L. K. & Lane, C. W. Treatment of periungual warts. *Missouri Medicine*, 1953, 50, 765.
4. Harvey, K. M. Pigmented naevus of nail. *Lancet*, 1960, ii, 848.
5. Kaminsky, C. A., de Daminsky, A. R., Shaw, M., Formentini, E. & Albulafia, J. Squamous cell carcinoma of the nail bed. *Dermatologica*, 1978, 157(1), 48.
6. Lamp, J. C., Graham, J. H., Urbach, F. & Burgoon, F. Jr. Keratoacanthoma of the subungual region. *Journal of Bone and Joint Surgery*, 1964, 46A, 1721.
7. Leppard, B., Sanderson, K. V. & Behan, F. Subungual malignant melanoma: Difficulty in diagnosis. *British Medical Journal*, 1974, 1, 310.
8. Lewing, K. Subungual epidermoid inclusions. *British Journal of Dermatology*, 1969, 81, 671.
9. Stoll, D. M. & Ackerman, A. B. Subungual keratoacanthoma. *American Journal of Dermatopathology*, 1980, 2(3), 265.
10. Undeutsch, W. & Shrieferstein, G. Garlic corm fibroma. *Dermatologica*, 1974, 149, 110.
11. Yung, C. W. Subungual epidermal cyst. *Journal of American Academy of Dermatology*, 1980, 3(6), 599.

Chapter 16

Toenails

Reviews on the fingernails do not always include the toenails. Certainly the changes in the toenails usually accurately reflect the nail plate changes in the fingers. The demands of teachers of podiatry, the generally aging population and the continuing demands of high fashion shoes have led to an increasing interest in examination of the toenails. The examples which follow are those seen in general internal medicine practice and are also used during tutorial courses at the New Zealand School of Podiatry.

Onychomycoses and ingrowing nails provide many referrals in the younger and middle-aged groups and peripheral vascular disease and onychogryphosis in the elderly. Above all, the toenails, their transverse ridging, rounding and frequent mycotic infections reflect the effect of footware. Inadequately designed or unsuitable shoes and boots provide many of the changes seen in general family practice.

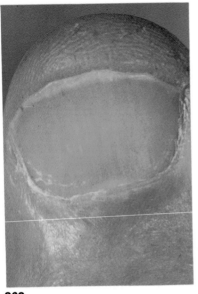

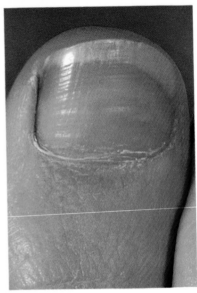

363   **364**

363 Normal. Healthy great toenail from a young female.

364 Normal – some transverse lines. Normal great toenail from a young woman who habitually wears open-toed sandals.

365 Normal. Great toenail from normal young female given to wearing tight 'court' shoes.

366 Normal variations. Great toenail from normal young female. Note small area of distal onycholysis and some 'ingrowing area on right margin'.

367 'Splitting'. ?Nutritional factor. Splitting of great toenail secondary to repeated mild trauma.

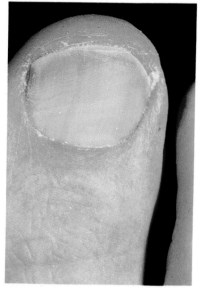

365

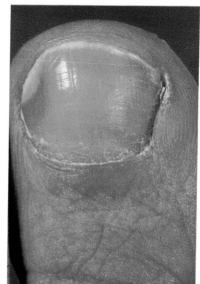

366

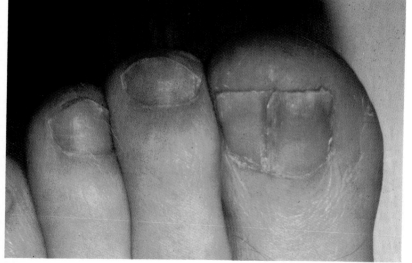

367

368

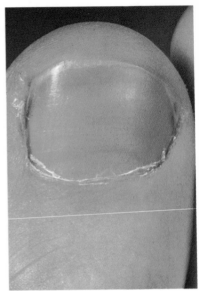
369

368 Healthy normal. Normal great toenail from young female. Some transverse lines suggesting episodic wearing of tight shoes.

369 Normal. Normal great toenail from a young female scientist.

370 Normal. Normal great toenail from young receptionist who has excessively tight shoes.

371 Congenital onychodysplasia toenail. Additional nails are more common in the fingers. Reports to date usually involve the index finger. Here in an intellectually handicapped young woman with normal fingernails. These small abnormal nails occur in the foot.

372 Normal toenail. Normal toenail of a male clerical worker in his 30s. Note the tendency for lateral groove to be grown over by eponychium.

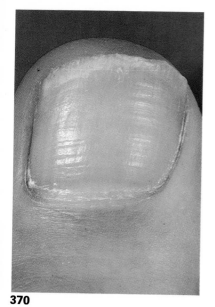

370

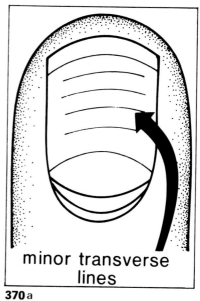

minor transverse
lines

370a

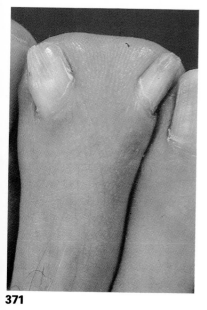

371

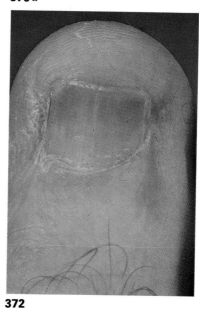

372

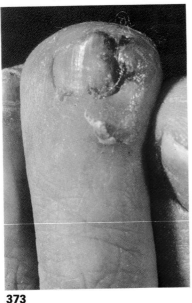
373

374

373 Unsuspected diabetes. Toe infection with nail damage secondary to long-standing but undiagnosed and untreated diabetes mellitus.

374 Ischaemia. Great toenail from man in his 60s with one leg amputated for vascular disease and second leg endangered. Note colour of skin which also has a 'waxy' appearance.

375 Fungus. Onychomycosis of great toe (*Trichophyton*)showing lysis and thickening.

376 Fungus *(scopulariopsis)* in several toenails, confirmed by scrapings in a young man working in gumboots.

377 Infection. Excision of nail secondary to gross staphylococcal infection.

378 Great toe tinea. Onychomycosis or fungal infection of great toenail giving rise to dystrophy.

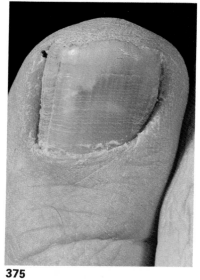

375

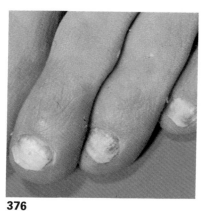

376

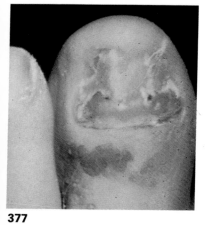

377

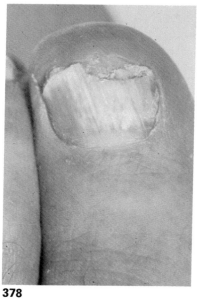

378

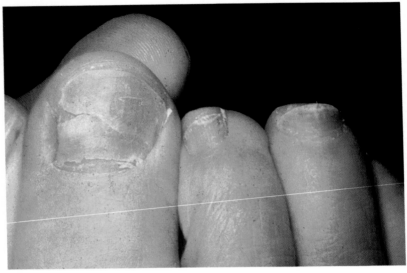

379

379 Raynaud's. 'Shredding' of great toe in an elderly patient with Raynaud's disease.

380 Ischaemia rather than tinea. Nail dystrophy due to ischaemia of foot. Male, 68, long-term smoker with no run-off in angiogram. Note reddening of skin and scaliness of skin.

381 Onychorrhexis of great toenails.

382 Pancreatic disease. Changes in great toe resulting from cystic fibrosis. No tinea found and no other skin lesions. Mechanism unknown.

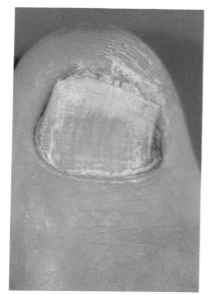

380

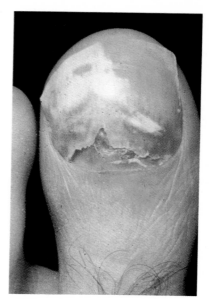

382

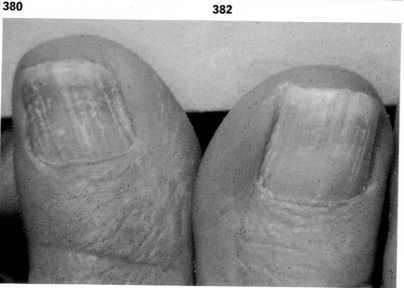

381

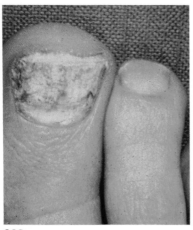

383

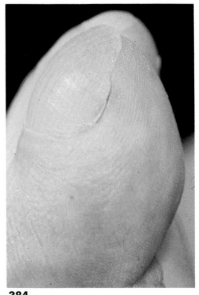

384

383 Onychomycosis with lifting and loss of toenails.

384 Clubbing of toes. Great toenails lateral view of marked clubbing. Note lattice work of longitudinal ridges and some transverse lines.

385 'Greenness' due to trauma. Trauma to great toenail (weight dropped on toe) leading to subungual haematoma.

386 Yellow nail syndrome: toenail involvement. Yellow nail syndrome in a man in his 60s who presented with hydrothorax and some swelling in the leg. This suggests abnormalities in leg of lymphatics as well as in abdominal site.

387 Chronic fungus. Chronic fungal infection leading to secondary staphylococcal paronychia and nail dystrophia in a diabetic patient with four-fold above normal blood sugar levels.

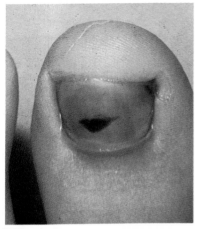

385

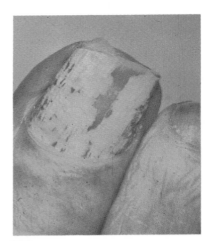

387

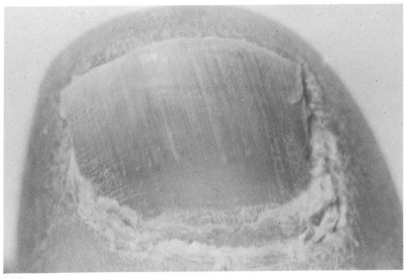

386

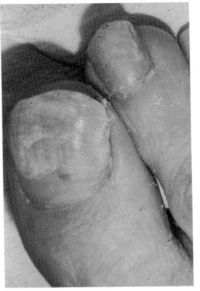
388

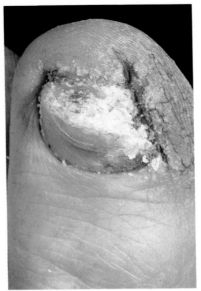
389

388 General skin disease. Eczema of nails – part of a generalised eczema.

389 Treated onychogryphosis of great toe. Note tissue injury to soft tissue.

390 Trauma. Splinter under great toenail.

391 Mild yellow nail syndrome. Mild appearances of yellow nail syndrome in toenails of a 56-year-old man with life-time history of congenital lymphatic abnormalities in the legs.

390

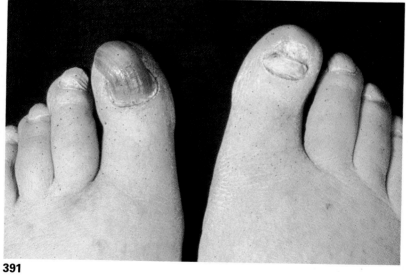

391

231

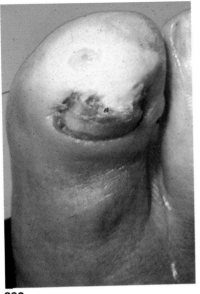

392

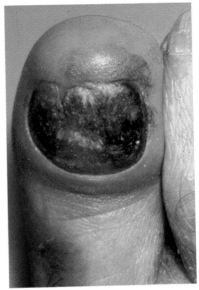

393

392 Peripheral neuropathy. Ulceration of tip of great toe secondary to peripheral neuropathy.

393 Secondary paronychia. Acute paronychia in great toenail subsequent to trimming of an onychogryphosis.

394 Onychogryphosis of great toe.

395 Ischaemia. 71-year-old man with heavy smoking history complaining of cold feet and poor eyesight (early cataract). Attempted home treatment of 'ingrowing' toenail with staphylococcal infection and necrosis in sullus.

396 Treatment of onychogryphosis. Severe onychogryphosis of great toenail in elderly man with ischaemia of the foot – diabetes and some secondary paronychia. First phase of treatment by reduction of nail and clearing up infection.

397 A unicorn!

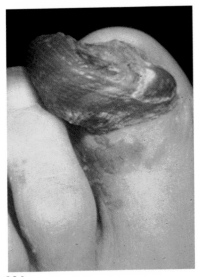

394

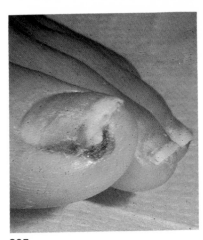

395

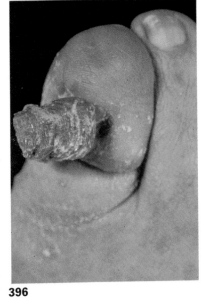

396

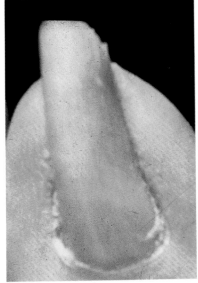

397

233

398

398 Onychogryphosis. Note the tendency to a greenish-yellow colour in the abnormal nail. Skin also shows signs of ischaemia.

399 Onychogryphosis. Urgent treatment is required for the second toenail deformed by tight shoes. The scaliness of the skin is thought to be due to ischaemia and was shown to be the result of thiamine deficiency or beri beri.

400 Onychogryphosis. Again shows unusual discolouration in the great toe (compare with second toe). Both toe nails need treatment but care will be required because of evidence of ischaemia as shown by shiny and scaly skin.

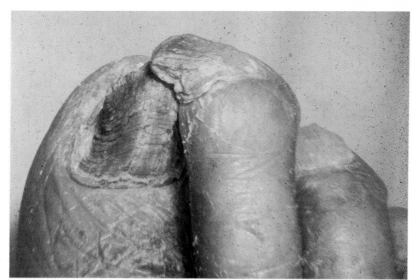

399

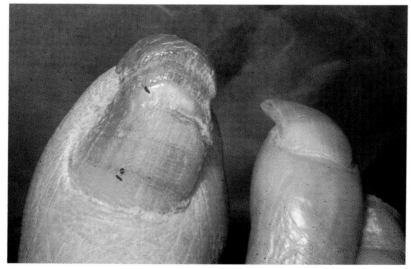

400

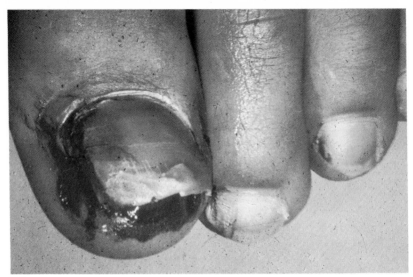

401

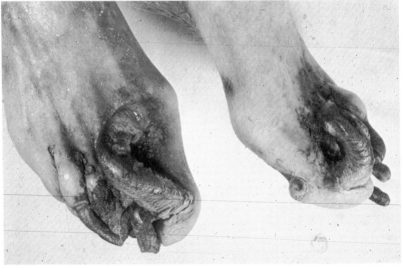

402

236

401 Trauma. Avulsed nail due to trauma.

402 Extensive onychogryphosis. Very severe onychogryphosis in a demented old lady living on her own in an isolated dwelling.

References

1. Baran, R. & Badillet, G. Primary onycholysis of the big toenails: a review of 113 cases. *British Journal of Dermatology*, 1982, **106**(5), 529.
2. Brereton, R. J. Simple old surgery for juvenile embedded toenails. *Zeitschrift fur Kinderchirurgie und Grenzgebiete*, 1980, **30**(3), 258.
3. Muto, H. & Yoshioka, I. Relationship between the degree of coverage of the nail root by the posterior nail wall and the length of the visible part of the nail in human toes. *Kaibogaku Zasshi*, 1977, **52**(4), 269.
4. Parker, S. G. & Diffey, B. L. The transmission of optical radiation through human nails. *British Journal of Dermatology*, 1983, **108**(1), 11.
5. Robb, J. E. Surgical treatment of ingrowing toenails in infancy and childhood. *Zetischift fur Kinderchirurgie und Grenzgebiete*, 1982, **36**(2), 63.
6. Siegle, R. J. & Swanson, N. A. Nail surgery, a review. *Journal of Dermatological and Surgical Oncology*, 1982, **8**(8), 659.
7. Thomas, P. S. & Nevin, N. C. Radiological findings in the DOOR syndrome. *Annals of Radiology (Paris)*, 1982, **25**(1), 54.

Index

Leukonychia : 8° to injury → next Step → Beau line
Kertes Nail & lunule & Surrounding Tissue.

Long. Ridges

a normal Variation in middle life (>40y/o)
mild trauma Ca Produce → mild onycholysis
Beaus lines are "growth arrest line"

Koilonychia

1. Fe^{++} def   2. ↓ Protein (A⁴) enables)   3. Chronic D.M

cuneiformis def

1. NUTR. problems   2. Trauma (central) to matrix

Art. ind. c f . . . .
Trauma A

Onycholysis

Trauma, thyroid, Psoriasis
Photo, Contact, drug

R10 L.P.

Terry's ½ / ½   (P.152)

A ↓LFT (↓ALB.)
B. Muehrcke's lines (↓↓LFT)
C. ↓ renal

ridges can be 2° to cuticle trauma (NOT a
    Beau line)

Normal Variation

1. long. ridges   2. beading 3. long. ridges
9. Trans. lines